SCIENTIFIC REALISM AND THE RATIONALITY OF SCIENCE

Scientific Realism and the Rationality of Science

HOWARD SANKEY
University of Melbourne, Australia

Routledge
Taylor & Francis Group
LONDON AND NEW YORK

First published 2008 by Ashgate Publishing

Published 2016 by Routledge
2 Park Square, Milton Park, Abingdon, Oxon OX14 4RN
711 Third Avenue, New York, NY 10017, USA

Routledge is an imprint of the Taylor & Francis Group, an informa business

British Library Cataloguing in Publication Data
Sankey, Howard
Scientific realism and the rationality of science
1. Realism 2. Science – Philosophy
I. Title
501

Library of Congress Cataloging-in-Publication Data
Sankey, Howard.
Scientific realism and the rationality of science / Howard Sankey.
p. cm.
Includes bibliographical references and index.
ISBN 978-0-7546-5888-7 (hardcover : alk. paper) 1. Realism. 2. Science—Philosophy. I. Title.

Q175.32.R42 S26 2007
501—dc22

2007007959

ISBN 9780754658887 (hbk)
ISBN 9781138251823 (pbk)

Contents

Acknowledgements

The papers assembled here represent work I have undertaken in recent years on issues relating to scientific realism. While engaged in this work, I have benefited from the support of a number of institutions. The Department of History and Philosophy of Science at the University of Melbourne has provided me with my academic home base throughout the time that I have been engaged on these topics. I was fortunate, once again, to enjoy the hospitality and stimulation of the Center for Philosophy of Science at the University of Pittsburgh, where I held a visiting fellowship during the Fall semester of 1998. On two occasions, in the first half of 1999 and the latter part of 2002, I received a warm reception as Guest Professor at the Center for Philosophy and Ethics of Science at the University of Hanover. One paper even began life while I was still affiliated with the Philosophy Department of Saint David's University College, Lampeter in 1991. Finishing touches were applied to several of the essays while I was a visitor at the Poincaré Archives at the University of Nancy. I would like to take this opportunity to express my gratitude to these institutions for their support, both material and intellectual.

Much of the material contained in this book has served as the basis for presentations which I have given at a variety of locations. Apart from talks given at the University of Melbourne, the work included here formed the basis of seminars and lectures presented in a variety of seminar series and conferences held at the Universities of Bielefeld, Brasilia, Genoa, Hanover, Pittsburgh and Roskilde, the Catholic University of Louvain, La Trobe University, the Humboldt University of Berlin, Saint David's University College, Swarthmore College, and the Belgian Society for Logic and Philosophy of Science in Brussels. I also gave papers based on this work at annual meetings of the Australasian Association of Philosophy in Melbourne in 1999 and the New Zealand division in Dunedin in 2005, as well as the Australasian Association for the History, Philosophy and Social Studies of Science in Melbourne in 2001 and 2003. I am grateful to the organizers of these symposia for the invitations to present my work, as well as to audience members for the opportunity to discuss my ideas with them. The comments received in such settings are a valuable source of stimulation which often suggest new directions and enable me to see issues in different ways.

As always, I am deeply indebted to philosophical friends and colleagues, with whom I have discussed these topics, and who have provided me with comments on my work. I would like to take this opportunity to express my gratitude to the following for exchanges in relation to the topics dealt with in this book: Paulo Abrantes, Martin Carrier, Steve Clarke, John Collier, David Cockburn, Michael Devitt, Brian Ellis, Michel Ghins, Dimitri Ginev, Paul Hoyningen-Huene, Hugh Lacey, Larry Laudan, Tim Lyons, Michele Marsonet, Alan Musgrave, Robert Nola, Stathis Psillos, Nicholas Rescher and Harvey Siegel. Without such interlocutors, philosophy would be a solitary game indeed.

The papers which make up the content of this book have all been published previously. Three of the chapters incorporate material that was originally published in separate papers. The remainder have been revised in order to improve clarity of expression, extend arguments, or reduce repetition of similar content.

I would like to thank the editors and publishers of the following publications for permission to reproduce this material here.

Chapter 1, 'Scientific Realism', consists of material drawn from two articles: 'What is Scientific Realism?', *Divinatio* 12 (2000), 103-20 and 'Scientific Realism: An Elaboration And A Defence', *Theoria* 98 (2001), 35-54.

Chapter 2, 'The God's Eye Point of View', was originally published as 'Scientific Realism and the God's Eye Point of View' in *Epistemologia* XXVII (2004), 211-26. It includes as an appendix a section from 'Realism Without Limits', *Divinatio* 20 (2004), 145-65.

Chapter 3, 'Truth and Entity Realism', was originally published as 'The Semantic Stance of Scientific Entity Realism' in *Philosophia* 24: 3-4 (1995), 405-15.

Chapter 4, 'Incommensurability and the Language of Science', combines material which originally appeared in 'The Language of Science: Meaning Variance and Theory Comparison', *Language Sciences* 22 (2000), 117-136 and 'Incommensurability: The Current State of Play', *Theoria* 12: 3 (1997), 425-45.

Chapter 5, 'Induction and Natural Kinds', was originally published in *Principia* 1: 2 (1997), 239-54.

Chapter 6, 'Methodological Pluralism, Normative Naturalism and the Realist Aim of Science', was originally published in R. Nola and H. Sankey (eds), *After Popper, Kuhn and Feyerabend: Recent Issues in Theories of Scientific Method, Australasian Studies in History and Philosophy of Science, Volume 15*, (Dordrecht: Kluwer Academic Publishers, 2000), 211-29.

Chapter 7, 'Realism, Method and Truth', was originally published in M. Marsonet (ed.), *The Problem of Realism* (Aldershot: Ashgate, 2002), 64-81.

Chapter 8, 'Why is it Rational to Believe Scientific Theories are True?', was originally published in C. Cheyne and J. Worrall (eds), *Rationality and Reality: Conversations with Alan Musgrave*, (Dordrecht: Springer, 2006), 109-32.

Introduction

The aim of this book is to articulate and defend a scientific realist philosophy of science. Throughout my philosophical work, I have adopted a thoroughly realist outlook. This was an outlook I acquired in the course of my initial training in philosophy in New Zealand. It was then reinforced by immersion in the Australian philosophical scene. But until recent years I have not overtly sought to defend the position of realism.

The situation began to change not long after the publication of my first book, *The Incommensurability Thesis* (1994). In that book, I sought to show that the semantic variance which lies behind claims of semantic incommensurability may be analyzed in an unproblematic manner within a scientific realist framework.[1] I employed a modified causal theory of reference to argue that translation may fail between the special vocabulary of scientific theories, while sufficient commonality of reference is maintained to ensure that the content of the theories may be compared. Thus, the occurrence of profound conceptual change within science poses little threat to either a realist view of scientific progress or a rationalist view of scientific theory-choice.

Following publication of *The Incommensurability Thesis*, my philosophical interactions began to take place in a more global arena. As a result, I came into sustained contact with philosophers of science working in continental Europe, for whom scientific realism was a profoundly problematic position, if not simply anathema. While I had sought to show that a basis for theory comparison exists within the sciences, little common ground was afforded by realism in these discussions. Indeed, I found realism to be a decidedly unpopular and poorly understood position, all too easily dismissed as a naive doctrine subject to decisive objections. I soon came to realize that realism requires careful elaboration and defence, if it is to be made plausible to philosophers schooled in opposing philosophical traditions.

That is one of the tasks that I have undertaken in several of the essays that are included in the present book. At the same time, this book is an attempt to make good on a promissory note issued in the Introduction to my second book, *Rationality, Relativism and Incommensurability* (1997). That book too dealt with aspects of the problem of semantic incommensurability, while also addressing the question of rational theory-choice in the context of variant standards of theory-appraisal. While realism figured only marginally as a topic in *Rationality, Relativism and Incommensurability*, the views of conceptual change and rationality which I presented there fit comfortably into a realist framework. At the end of the Introduction I wrote that I hoped soon to provide an elaboration of the connection between realism and the

1 I distinguish semantic incommensurability, which arises due to semantic variation between theories, from methodological forms of incommensurability, which stem from variation in methodological standards of theory-evaluation (see Sankey and Hoyningen-Huene, 2001, ix).

views of conceptual change and rationality that I presented in that book. In various ways, the essays collected in the present book all seek to deliver on that promise.

My main reason for not dealing with realism as a central topic of *Rationality, Relativism and Incommensurability* was that I then regarded the issues of conceptual change and theory-choice to be distinct issues from the issue of realism.[2] This is perhaps the nub of the issue. For the realist, the world we inhabit, and which science investigates, is an objective reality whose existence and character are independent of human thought. Truth, too, is objective, since it is a non-epistemic relation of correspondence between language and the mind-independent world. Neither variation of concepts nor choice of theory impinges on the nature of the objective world or on the truth about that world. In this sense, the issues of conceptual change and rational theory-choice are indeed distinct from the issue of realism.

But it was just this separation of realism from conceptual and epistemic issues which was placed in dispute in my exchanges with European philosophers. The issue was often expressed in terms of the relationship between epistemology and ontology. 'Since Kant', I was told, 'epistemology and ontology must go hand-in-hand.'[3] Sometimes, the point was the simple one that one must reflect upon the epistemic basis of one's views about the world rather than dogmatically assert the world to be a given way. At other times, the point was the stronger one that the world of which we have knowledge is in part constituted by our conceptual and epistemic activities. On still other occasions, I was told that we must do away with the entire distinction between subject and object on which the distinction between epistemology and ontology is based.

As a realist, I hold that the way the world is does not depend on what we believe about it or how we conceive of it. In that sense, epistemology and ontology are distinct. But I have come to agree in part with my European colleagues that epistemology must go hand-in-hand with ontology. Given the realist insistence on the mind-independence of reality, human cognitive activity is not constitutive of reality. There is a gap between mind and world. This raises the question of how knowledge (scientific or otherwise) is possible. The realist must explain how human cognitive activity gives rise to genuine knowledge of a mind-independent reality.

While epistemology and ontology are distinct in the sense that reality is independent of thought, they must be brought into a relation of mutual support within the context of a realist philosophy of science. On the one hand, scientific realism requires an epistemology. For it must be shown how the methods of science are able to produce knowledge of a world that exists outside the mind. On the other

2 It is also the case that the defence of untranslatability, methodological pluralism and a non-algorithmic conception of rationality, which was the principal aim of *Rationality, Relativism and Incommensurability*, is neutral with respect to the question of scientific realism. One need not be a realist to endorse these views.

3 For example, in an interview which I conducted with Paul Hoyningen-Huene in 1996, he commented that: 'I think since Kant in philosophy in general it has been on the agenda that epistemological questions cannot be discussed apart from metaphysical questions, and *vice versa*. I mean it doesn't make sense to speak about metaphysics without questioning our possibilities of access to the things which there are' (Sankey and Hoyningen-Huene, 1996, 61).

hand, the epistemology of realism requires support from ontology. For in order to explain how the methods of science produce knowledge, it must be explained how the world is such that it is accessible by such means. Thus, epistemology and ontology do go hand-in-hand, but not in the sense that knowledge and world are mutually constitutive.

As for how epistemology and ontology are to be conjoined, here I draw on a theme that came to the fore in the final chapters of *Rationality, Relativism and Incommensurability*. There I advocated a turn toward naturalism in the philosophy of method. In particular, I sought to show how a normative naturalist account of epistemic warrant of the kind advocated by Larry Laudan may be employed to meet the challenge of relativism. Unlike Laudan, though, I hold that the normative naturalist account of warrant is able to serve the purposes of a realist epistemology of science. I agree with Laudan that the methods of science may be regarded as tools of inquiry that are employed in pursuit of epistemic goals. I agree too that the warrant for such methods rests on their historical track record in furthering such goals. Where I depart from Laudan is in arguing that the best explanation of the success of theories produced on the basis of these methods is that the methods are in fact a reliable guide to truth. Epistemology and ontology are thus conjoined because the world we inhabit is one in which the methods of science are a reliable means of securing knowledge about it.

With the foregoing as background, I will now present an overview of the essays that follow. The first priority is to formulate scientific realism in as clear a manner as possible. Hence, Chapter 1 spells out the scientific realist position as I understand it, and presents the basic arguments on its behalf. I propose what I take to be a standard construal of the scientific realist position as a form of the traditional metaphysical realist doctrine that the world exists independently of the mental. The realist position is a position of epistemic optimism, which holds against the sceptic that humans are able to acquire knowledge of the world. The specifically scientific realist dimension arises because such knowledge extends to unobservable aspects of the world investigated by theoretical science. For the realist takes scientific progress to consist in advance on truth about both observable and unobservable aspects of the world. Truth is understood, as indicated previously, as a correspondence relation. Apart from these basic elements of the scientific realist position, a number of further semantic and metaphysical tenets will also be discussed, though I regard them as optional extras which are not essential to the scientific realist position.

As for the arguments for scientific realism, here it is important that scientific realism be understood as a form of realism in general. For a number of powerful arguments which provide groundwork for the case for scientific realism are in fact arguments for realism in general. There are two such arguments. The first is an argument for realism as the position which best reflects a non-anthropocentric view of our place in the natural world. The second is an argument for realism at the commonsense level, which treats realism about unobservable entities as an outgrowth of commonsense realism. Such arguments provide the groundwork for scientific realism. Once the general realist outlook is established, one may then proceed to argue for realism about science. Here, too, my approach tends to follow realist orthodoxy. The basic argument at this level turns on the success of science. It is the argument that scientific realism is the best explanation of the success of science. I shall present this argument at two levels. First, at the level of theories, I argue that the success of science is best explained by the truth or approximate truth of theories. Second, at the level of method, I argue that the success of theories produced by the methods of science is best explained by the truth-conducive character of such methods.

One of the major concerns with scientific realism that I have encountered relates to the perspective occupied by the philosopher who adopts the realist stance. In proposing that humans acquire knowledge of a mind-independent world, it would appear that the realist must adopt a perspective that is situated outside the human perspective. For in order to be able to judge that human mental states are true representations of such a world, it must be possible to adopt an objective stance from which to compare the human perspective with reality. That stance must surely lie outside the human perspective. In the words of Hilary Putnam, it must be a God's Eye point of view. Since no human could possibly occupy such a perspective, and the realist position requires such a perspective, the realist position may be rejected as untenable.

In Chapter 2, I consider the objection from the God's Eye point of view. I argue, in the first place, that it is not necessary for the realist to adopt a God's Eye point of view, since realism may be proposed from within our human perspective as a hypothesis about the relation between mind and reality. However, I also attempt to show that there is a perfectly intelligible sense in which we are able to adopt an external perspective. This may be seen by reflecting in a naturalistic manner on the scientific study of animal cognition. Scientists are able to investigate the epistemic relations which non-human animals bear to reality. But there is no reason why a similar investigation may not be undertaken of our own epistemic relations to the world. Just as we may study animal cognition, so, too, we may investigate the relations between our own minds and the world. Thus, even if realism were to require an external standpoint, this is no basis for an objection to realism.

The notion of truth plays an important role in scientific realism on my construal of the doctrine. For the aim of science is to discover the truth, and thereby to advance our knowledge of the world. Thus the form of scientific realism which I propose differs from a version of scientific realism that has been advocated by a number of influential realist authors. Michael Devitt, Brian Ellis, Ian Hacking, and others, have argued for an ontological version of scientific realism that is known as entity realism. Entity realists eschew or downplay the notion of truth, emphasizing instead the reality of the unobservable entities discovered by science. By contrast with entity realism, the form of scientific realism presented in this book constitutes a semantic version of realism in Michael Devitt's sense, since it makes use of the notion of truth (e.g., Devitt, 1991).

There are reasons to find the stance of entity realism appealing. Avoidance of the notion of truth is attractive, since it may enable objections which relate to the notion of truth to be evaded. Moreover, the emphasis on entities rather than theories promotes the study of experimental practice, which is a significant corrective to traditional emphasis on theoretical science. Still, for a number of reasons, I favour the more standard, truth-orientated conception of scientific realism described above. For one thing, a scientific entity realism which eschews semantic notions such as truth seems to assume the possibility of referring to theoretical entities in the absence of individuating descriptions which are true of such entities. Such a position presupposes the viability of a pure causal theory of reference determination for theoretical terms, of a kind which seems quite implausible.[4] For another thing, there is the matter of semantic ascent. Assertion of the existence of an entity commits one, via semantic ascent, to the truth of at least the existence claims relating to the entity. So it is not clear that use of the notion of truth may be entirely avoided by the entity realist. Nor is it clear, therefore, that semantic issues may be evaded by the entity realist. Finally, one of the best-known arguments for entity realism is the so-called experimental argument for realism, which was originally presented as a distinct argument from the success argument for realism (Hacking, 1983, 271). But it seems clear that the experimental argument is simply a version of the success argument applied in the context of experimentation (Resnik, 1994). So it fails to import a new argumentative strategy into the realist's repertoire.

While I support a truth-orientated version of realism, it is important to emphasize that this indicates no less a commitment to realism about the mind-independent world than the commitment of the entity realist. Entity realists are right to insist that the fundamental commitment of realism is a metaphysical commitment to the existence of a mind-independent world, as well as to the various entities which are found to populate it. But such metaphysical commitment is fully compatible with endorsement of a semantic version of realism. Indeed, commitment to the mind-independent world and to the entities of science and common sense constitutes a major component of the truth-orientated version of scientific realism advocated here.

I am inclined to regard entity realism as a version of scientific realism. The entity realist is a close ally in disputes with a range of anti-realist critics of realism, such as Kantian constructivist positions, and epistemically sceptical positions. Still, the question arises of whether the position of entity realism is a genuine alternative to scientific realism. In Chapter 3, I examine the question of whether entity realism does constitute a distinct position from semantic formulations of scientific realism. I seek to show that it can indeed be legitimately distinguished from semantic versions of scientific realism. Entity realism does not express, nor does it immediately entail, a semantic thesis involving the notion of truth. While it is possible to derive a semantic thesis from entity realism by semantic ascent, the resulting semantic thesis is not committed to any particular theory of truth. So entity realism entails a semantic thesis in only an attenuated sense.

Semantic concerns are also the focus of Chapter 4, where I revisit the problem of incommensurability. The phenomenon of conceptual change or meaning variance in science raises a fundamental difficulty for the realist in relation to the progress of science. Given the realist view that scientific progress consists in advance on truth, such progress requires an increase in truth known about the world. The connection between truth and reference is therefore of crucial relevance to the issue of progress. For in order for the transition between theories to constitute progress in a realist

4 For discussion of problems relating to the reference of theoretical terms, see my (1994, Ch. 2).

sense, a later theory must yield an increase in the truth known about the same entities as those investigated by earlier theories. With the exception of cases of radically mistaken ontology, a later theory must refer to the same set of entities as earlier theories referred to. Otherwise it is not possible for the transition between the theories to constitute progress toward truth in the same domain of inquiry.

The problem of semantic incommensurability is a major problem for a realist account of scientific progress precisely because it casts doubt on the continuity of reference between theories. For if theoretical change involves significant change in the concepts expressed by scientific terms, then there may be a discontinuity of reference in the transition between theories. The problem of stability of reference between meaning variant theories is at base a problem in the theory of reference about the nature of reference determination. This was a major topic of *The Incommensurability Thesis*. For detailed coverage of the topic, I refer the reader to that book.

In Chapter 4, I present an overview of the emergence of the problem of incommensurability within the context of 20th century philosophical thinking about the language of science. The chapter traces discussion of the language of science from the verificationism of the early logical positivists through to the partial interpretation model of later positivism and the theoretical context account of meaning favoured by post-positivist philosophers of science of the historical school. It then considers the problem of reference change in light of alternative (descriptive, causal) models of reference determination, before presenting the causal-descriptive model for which I argued in *The Incommensurability Thesis*.

While realism is an underlying theme of the chapter, it is the principal focus of the final two sections. The implications with respect to realism of the taxonomic incommensurability thesis proposed by Kuhn in his later work are examined, and it is argued that such incommensurability poses no threat to a realist philosophy of science. The chapter closes by rebutting the claim of Paul Hoyningen-Huene, Eric Oberheim and Hanne Andersen (1996) that my causal-descriptive approach to the problem of incommensurability presupposes realism and thereby succumbs to a meta-level incommensurability with anti-realist proponents of the incommensurability of scientific theories.

From semantic aspects of realism, attention then shifts to concerns of an epistemological nature. Throughout the four final chapters of the book, I adopt a broadly naturalistic approach to the epistemological questions under consideration. My principal aim is to demonstrate that the normative naturalist account of epistemic warrant may be embraced within a realist framework, and to employ this account of warrant as part of my argument for a realist epistemology of science. In the process, I explore the idea that the epistemology of scientific realism must draw on considerations of a metaphysical nature.This exploration commences in Chapter 5, which illustrates the epistemological relevance of metaphysical considerations in the context of the traditional Humean problem of induction. The leading idea of the chapter is a familiar one from the history of attempts to solve the problem of induction. It is the idea that what ultimately provides the rationale for our use of induction is the fact that the world is an ordered reality governed by underlying laws. This is a version of the principle of the uniformity of nature. Philosophers have usually rejected the principle of uniformity as a solution to the problem of induction because it is unable to be established independently of induction. It cannot, therefore, serve as justification of induction.

But I think that such a rejection of the principle of uniformity of nature is ill-advised for the realist. Rather than reject the principle out of hand, the realist may instead employ realist metaphysical commitments as the underpinnings for an epistemology. This will require the realist to articulate a metaphysical position that goes beyond a minimal commitment to an objective reality. In certain contexts, it is useful to characterize realism (as I do in Chapter 1) as committed to the existence of a mind-independent reality, for example, in order to distinguish realism from a Kantian or idealist position. But a mind-independent reality may be an amorphous, unordered world. Such a world is hardly a world worth fighting for. The world in which the realist should believe is not just a mind-independent reality, though it is at least that. It is a world with structure and order. But the existence of such structure and order has epistemological implications. Recognizing this is crucial to the development of a realist epistemology.

Though the argument of Chapter 5 is a version of the traditional appeal to the uniformity of nature, it is not couched in the usual way as a blanket resemblance of past and future. Rather, I understand the uniformity of nature in terms of the operation of laws of nature which it is the task of science to discover. Following Brian Ellis, I see a close connection between laws of nature and the natural kinds with which our world is populated. Laws of nature are grounded in the irreducible causal powers of things that characterize members of a given natural kind. So, rather than understand the uniformity of nature as some sort of general resemblance of past and future, I see it as residing in the inbuilt behavioural tendencies of individual members of natural kinds.

This line of argument derives from the position proposed by Hilary Kornblith in his book, *Inductive Inference and its Natural Ground.* Kornblith argues that reliable inductive inference is grounded in the existence of natural kinds whose members share sets of homeostatically clustered properties in common. I depart from Kornblith only on points of detail. Where Kornblith chooses in naturalistic vein not to directly address Humean scepticism, I suggest instead that the appeal to natural kinds can be used to justify induction against the Humean sceptic. I also espouse the more substantive metaphysics of Brian Ellis's scientific essentialist theory of natural kinds, rather than rest content with the homeostatic property cluster model favoured by Kornblith.

It will not escape notice that this attempt to solve the problem of induction draws on metaphysical views of the kind that are set aside as optional doctrines of realism in Chapter 1. A brief word of explanation is therefore in order. My aim in Chapter 1 is to articulate the position of scientific realism in a manner that reflects the standard understanding of the position construed in a generic fashion. As it happens, the position of scientific realism presented in Chapter 1 is the general version of scientific realism that I espouse. However, the formulation of the realist position in Chapter 1 is intended to serve as a generic statement of the position that is capable of embracing paradigmatic examples of scientific realism, while at the same time allowing for variation amongst diverse realists on non-essential matters.

As the book progresses, my attention turns to the more specific task of developing an epistemology for realism. Thus, increasingly, the position presented is my own specific version of scientific realism, rather than a generic form of the doctrine, such as that propounded at the start of the book.

Kornblith's approach to the problem of induction is based on a thorough-going epistemological naturalism. It is antisceptical. It draws on empirical claims from and about natural science in establishing a metaphysical framework within which to treat epistemological questions. It employs research in cognitive psychology in approaching conceptual and inferential aspects of knowledge-acquisition. In my view, such a naturalistic approach to epistemological questions is a model of how the realist should proceed in developing a realist epistemology for science.[5]

Because of the non-epistemic character of correspondence truth and the mind-independence of reality, the realist must explain how use of the methods of science yields knowledge. Against those who deny the possibility of a realist epistemology, the naturalistic realist may treat the problem of knowledge as the broadly empirical problem of explaining how cognitive agents embedded in the natural world are able to use their epistemic capacities to promote their survival. The success of practical activity based on common sense and scientific exercise of our epistemic capacities serves as robust confirmation that such knowledge is not only possible but actual.

The main outlines of the naturalistic epistemology that I favour are presented in Chapter 6. As previously mentioned, I adopt a realist version of Larry Laudan's normative naturalist account of the warrant of the rules of method. According to normative naturalism, the rules of method are tools of scientific inquiry, which may be evaluated on the basis of their historical track record in securing the cognitive aims of science. The most widely attested strength of this approach is its ability to serve as a counter to epistemological relativism (see *Rationality, Relativism and Incommensurability*, Chapter 10). Given Laudan's opposition to realism, it is less widely appreciated that normative naturalism may be used in support of a realist epistemology of science. The aim of Chapter 6 is to show that normative naturalism is able to be incorporated within a scientific realist framework. I seek to show, *pace* Laudan, that it is possible to have knowledge at the theoretical level. I argue that it may be rational to pursue truth as an ideal, even if it is unattainable. And I attempt to show, again *pace* Laudan, that we are able to monitor our pursuit of truth because satisfaction of the rules of method may serve as a fallible indication of our progress toward that aim.

The fundamental epistemological problem of realism is the problem of establishing a connection between epistemic methods and non-epistemic truth. In Chapter 7, I explicitly confront this problem, which I refer to as the problem of method and truth. Those anti-realists who take truth or reality to depend on epistemic activity resolve the problem by treating truth as a product internal to the application of method. Those anti-realists who are sceptics about theoretical science deny that the problem may be solved since they deny that a connection may be established between method and truth. By contrast with anti-realists of either variety, I seek to

---

5 A similar naturalistic program for realism is found in Devitt (1991, 5.7-5.10; 2002, 22-5).

resolve the problem in the context of realism on the basis of a naturalistic account of the epistemic warrant of methodological rules. In particular, I argue that what best explains the use of the methods of science to produce successful theories is that the methods comprise a reliable means of securing the truth. The realistic version of normative naturalism proposed in Chapter 6 is the basis for the solution of the problem of method and truth that is presented in Chapter 7.

I continue to develop this approach to the problem of method and truth in Chapter 8, which is the final chapter of the book. I examine two solutions to the problem of method and truth that may be found in the work of one influential scientific realist, my former teacher, Alan Musgrave. Musgrave proposes two epistemic principles which indicate conditions under which it is rational to believe that a theory is true. The first is the principle that it is reasonable to believe the best explanation of a fact to be true. The second is that it is reasonable to believe the hypothesis which has best withstood criticism. While both principles highlight important principles of rational belief, I argue that in the case of neither principle is it explained why it is rational to believe that a theory is true, as opposed to, say, empirically adequate. To explain why it is rational to believe that a theory is true, it is not enough to enunciate a principle of pure epistemology. To explain why it is rational to believe that a claim is true of the mind-independent world something must be said about the world. Ontology must be conjoined with epistemology if the problem of method and truth is to be solved in a realist framework.

It may be seen from this overview that the essays contained in this book develop a number of interwoven themes. Because they were originally written for publication as separate articles, there is some overlap of content between some of the chapters of the book. However, in revising the papers for inclusion in the book, I have sought to remove excessive repetition of material. At the same time, the individual chapters remain self-contained, and may be read as stand-alone pieces.

The papers were originally prepared for a variety of different audiences with different backgrounds and levels of expertise. A number of the papers were consciously written in an attempt to bridge a communicative gap between philosophical traditions. Some were originally prepared for oral presentation, while others were conceived solely as written papers. The target audience of a number of the papers was envisioned to be specialists in the philosophy of science, while a number of others were directed at a non-specialist audience. While I have sought to maintain an even tone throughout the revised versions of these papers, it is unavoidable that the variation in original audience will have left its traces.

Chapter 1

Scientific Realism

1.1 Introduction

Tables and chairs, and people who sit at tables in chairs, are all objects composed of matter. Science tells us that the basic components of matter are atoms. Atoms themselves are made up of electrons, neutrons and protons. The neutrons and protons form a nucleus around which the electrons orbit. Apart from these particles, physicists have discovered numerous other particles, such as photons, quarks and neutrinos.

Unlike tables, chairs and people, the particles of which matter is composed are entities which cannot be directly observed using unaided sense perception. We do not come to know of the existence and nature of such unobservable entities by means of sense experience. Rather, our knowledge of unobservable entities is a matter of inference, conjecture and hypothesis. Scientists postulate the existence of such entities when they develop theories which explain the observed phenomena as the result of more basic occurrences at an unobservable level. Because scientists postulate unobservable entities within the context of scientific theories, philosophers of science refer to such entities as theoretical entities. The terms that scientists use to speak about such entities are often called theoretical terms to distinguish them from the observational terms that are used to speak about observable things.

When scientists tell us about such theoretical entities as atoms, electrons and other particles, what attitude should we adopt toward these claims? What is the status of such claims? Should they be treated as true, or approximately true, descriptions of actually existing things? Or are such claims to be treated as some sort of fictional discourse?

There is considerable disagreement among philosophers of science about the status of discourse about theoretical entities. Those who endorse the position of scientific realism say that scientific claims about theoretical entities should be taken literally, or at face value. They should be treated as true, approximately true, or at least as genuine attempts to make true or approximately true claims about objective reality. Those who endorse one or another anti-realist position reject the realist view that such claims should be regarded as true or approximately true claims about objective reality.

Anti-realist philosophers of science disagree among themselves about how theoretical claims are to be understood. One well-known anti-realist position is the position of instrumentalism, according to which talk of theoretical entities is no more than 'useful fiction' or a 'convenient shorthand'. But there are other anti-realist positions. According to internal realism, for instance, the theoretical claims of science are candidates for truth, but truth is relative to conceptual scheme or value system rather than correspondence to objective reality. By contrast, some contemporary versions of empiricism are sceptical with regard to theoretical knowledge. According to such sceptical anti-realism, theoretical claims about unobservable entities may well correspond to reality but empirical evidence may never be sufficient to provide support for the truth of such claims, which can at most be accepted as empirically adequate or warranted.

While there are many philosophers who find scientific realism a compelling and powerful doctrine, realism is not a view shared by all philosophers. There are many who fail to see the appeal of scientific realism. In part, this is because the doctrine of scientific realism is not always understood as well as it might be. In some cases, such lack of understanding may explain why its appeal is not noticed, though in many cases there are substantive disagreements which explain the lack of appeal. In this chapter, I will attempt to provide a clear presentation of scientific realism in the hope that such a presentation may broaden the appeal of the doctrine.

1.2 Scientific Realism as a Family of Doctrines

Scientific realism is not a simple thesis that may be embodied in any one single claim. Rather, scientific realism consists of a set of doctrines. Not all scientific realists agree about all of these doctrines. Thus, scientific realism is best characterized as a family of closely related doctrines.[1] Another way to put the point is to say that there is a variety of different versions of scientific realism. In this, of course, scientific realism is not alone. Anti-realism is also best thought of as a family of doctrines.

Despite being best characterized as a family of doctrines, some doctrines are more central to scientific realism than others. In what follows I will present six distinct doctrines which I take to form the core of scientific realism. There are, however, a number of other doctrines associated with scientific realism which should be mentioned as well. So I will also discuss several other doctrines which are often associated with scientific realism, but which may be treated as optional doctrines. Following the presentation of scientific realist doctrines, I will distinguish scientific realism from the doctrine of scientism, with which it is sometimes mistakenly identified. I will then present what I take to be the major arguments in favour of scientific realism.

Core Doctrines

The six doctrines which I will now present are doctrines which form the core of scientific realism. These doctrines are so central to scientific realism that a view which denies any of them can only lay claim to the title of 'scientific realism' in an attenuated sense. I present the doctrines in what I find to be a natural order of

---

1 For a sample of the varying characterizations of scientific realism found in the literature, see Devitt (1991, 98ff), Ellis (1990, 87-9), Hacking (1983, 21-31), Leplin (1984, 1-2), Newton-Smith (1981, 29, 38-9) and van Fraassen (1980, 8).

presentation. This order is not meant to suggest that any particular hierarchical or logical relation holds between the doctrines.

The first doctrine is a doctrine about the aim of science:

1. *Aim realism*: the aim of science is to discover the truth about the world, and scientific progress consists in advance toward that aim.

In the first instance, aim realism is a thesis about the aim of science. As such, however, it has immediate implications about the nature of scientific progress, which is why I formulate it in terms of both an aim and progress toward that aim. Because the aim of science is to obtain the truth, progress in science must consist in advance on that aim. Hence, scientific progress consists in advance on truth.

It is consistent with the doctrine of aim realism to deny that science has made much progress toward the aim of truth. Indeed, it is consistent with aim realism to deny that any progress at all has been made toward that aim. But scientific realists typically wish to say that a great deal of truth has already been discovered in at least some areas of science. Some may even be prepared to commit themselves to the truth of at least some of the well-established claims of the sciences.

However, along with most contemporary philosophers of science, scientific realists tend to understand science as an ongoing historical process that is, in all likelihood, far from complete. As a result, they do not assume that contemporary science has already achieved the aim of truth. At best, current scientific theories may be close to the truth, or they may be approximately true. Pursuit of the aim of science cannot, therefore, consist simply in the pursuit of the truth. It must also consist in seeking to advance toward the aim of truth by increasingly close approximation to that aim.

Three further points are worthy of note. First, the idea that truth is the aim of science is in need of further qualification. It is not just that science seeks truth. For there are endlessly many trivial truths of no particular interest to science. Rather, science seeks to discover truths which are particularly revealing and interesting. Since science seeks to explain phenomena, we may say that science seeks truths of an explanatory nature. Second, it is worth noting that the choice of the term 'discover' in the formulation of aim realism is a deliberate one. For the realist, the aim of science is to discover truth about the world. We do not invent, construct or fabricate the truth, as might be said by some anti-realist philosophers or by constructivist sociologists of science. Third, it is important to consider the status of the claim that truth is the aim of science. It is not an empirical hypothesis or generalization about the motivations of scientists, considered either as individuals or as a group. Nor is it a semantic claim to the effect that the concept of science is the concept of an enterprise that is directed toward truth. Rather, it is an epistemological claim that the purpose of a knowledge-seeking enterprise such as science is the pursuit of truth.

This latter point leads to the second core thesis of scientific realism. This is a thesis which relates to the nature of scientific knowledge:

2. *Epistemic realism:* scientific inquiry leads to genuine knowledge of both observable and unobservable aspects of the world.

The scientific realist does not simply assert that science seeks the truth about the world. Rather, scientific realism has a fundamentally epistemological rationale. For the scientific realist, the scientific pursuit of truth gives rise to genuine knowledge of the natural world. Scientific realism therefore entails epistemic realism, according to which scientific inquiry yields knowledge of the truth about the objective reality investigated by scientists.[2]

What most clearly distinguishes epistemic realism as a component of scientific realism is the insistence that scientific knowledge is not restricted to the observational level. It extends to unobservable aspects of reality as well. Epistemic realism is what characterizes scientific realism as an epistemological doctrine distinct from contemporary versions of empiricist philosophy of science which deny that it is possible to either have rationally justified belief or knowledge about unobservable states of affairs. It is also what epistemologically distinguishes scientific realism from neo-Kantian, constructivist views which deny epistemic access to the objective, mind-independent world, which lies beyond our phenomenal experience.[3]

The insistence that scientific knowledge extends beyond the observational level is further reflected in the third core claim of scientific realism. This is a thesis about the interpretation of theoretical discourse:

3. *Theoretical discourse realism*: scientific discourse about theoretical entities is to be interpreted in literal fashion as discourse which is genuinely committed to the existence of real unobservable entities.

Such realist treatment of theoretical discourse contrasts with an instrumentalist construal of such discourse. Instrumentalism denies the literal interpretation of theoretical discourse, treating it instead as fictional discourse. Theoretical entities are 'convenient fictions,' useful only as an aid to prediction. By contrast with instrumentalism, the scientific realist understands theoretical discourse to refer to events and regularities that take place at the unobservable level. Scientists explain observed phenomena on the basis of underlying causal processes. The explanations they provide refer to unobservable entities whose behaviour is responsible for the observed phenomena.

The realist treatment of theoretical discourse has both an ontological and a semantic dimension. At the ontological level, it is realistic about theoretical entities, since it implies that there really are unobservable entities which underlie observable phenomena. At the semantic level, it has general implications with respect to the reference of theoretical terms. For the point of a realist treatment of theoretical

2 The question arises of how precisely the notion of knowledge is to be understood in the context of scientific realism. As we will see in Chapter 6, I hold that something along the lines of the traditional justified true belief account of knowledge is a minimal condition for a realist conception of knowledge. This is particularly the case, given the need to produce a clear distinction between realism and assorted relativist and social constructivist conceptions of knowledge with which it contrasts.

3 The prime contemporary example of a neo-Kantian constructivist philosophy of science is, of course, Kuhn (1970a). For such an interpretation of Kuhn's metaphysical stance, see, for example, Devitt (1991) and Hoyningen-Huene (1993).

discourse is not just that theories are genuinely committed to the existence of theoretical entities. Rather, realist treatment assumes that theoretical terms may in fact succeed in referring to real theoretical entities. For the realist, the point of employing a theoretical term, such as 'electron', is specifically to refer to the theoretical entities in question, namely electrons. The realist therefore takes it to be possible for the theoretical terms employed by scientists to enter into relations of reference with the unobservable entities whose existence is postulated by theories.

The requirement of the possibility of reference does not commit the scientific realist to any particular theory of the reference of theoretical terms. At most, the possibility of such reference gives rise to a constraint on realistically acceptable theories of the reference of theoretical terms. But it does not entail any specific account of how the reference of theoretical terms is determined. Indeed, there is a range of theories of reference compatible with the requirement that theoretical terms be treated as genuinely referential expressions.[4]

The fourth core component of scientific realism is commitment to the basic metaphysical stance of realism about the external world:[5]

4. *Metaphysical realism*: the world investigated by science is an objective reality that exists independently of human thought.

Commitment to a mind-independent, objective reality is what most fundamentally characterizes scientific realism as a form of realism. On such a view, there is a world that exists independently of our thought, the existence, structure and features of which depend in no way on human experience, beliefs, concepts or language. It is a world of objects, properties, relations and facts, which we must discover by means of empirical inquiry. It is not a world which is in any way constituted or constructed out of the concepts or theories which we formulate as part of the process of empirically finding out about the world.

I employ the expression 'metaphysical realism' in the ordinary sense of commitment to the existence of a mind-independent reality. This use of the expression differs from the way it is employed by Hilary Putnam, who uses it to stand for the view that there is a fixed totality of mind-independent objects, determinate relations

---

4 The requirement of referential realism may be satisfied by a variety of theories of reference, which range from pure descriptive, to causal-descriptive and pure causal accounts. As it happens, I advocate a causal-descriptive theory of reference as part of the approach that I have proposed to the problem of the incommensurability of scientific theories (Sankey, 1994). However, no commitment to a specific account of reference is required by the realistic interpretation of theoretical discourse.

5 The expression 'external world' is the expression traditionally employed by philosophers to formulate the claim that there is a material world, which exists independently of the human mind. However, the expression itself is objectionable, since it seems to imply an untenable metaphysical divide between internal and external worlds, as well as to suggest that we are not part of the world. A further problem is that talk of an external world may provide the basis for the sceptical problematic – for example, Cartesian questions about the certainty of our knowledge of an external world – which should itself be rejected in favour of a naturalistic perspective which denies the legitimacy of such sceptical questions.

of reference between terms and objects, and one true, complete theory of the way these objects are (Putnam, 1981, 49). I prefer a more minimal characterization of metaphysical realism in terms of commitment to a mind-independent world. Putnam's use of the expression involves a number of controversial assumptions which the scientific realist need not embrace. For example, not all scientific realists may wish to assert either that there is a fixed totality of mind-independent objects or that it is possible in principle for there to be a single, complete theory of the world.[6]

The fifth component of scientific realism is a thesis about the nature of truth:

5. *Correspondence theory of truth*: truth consists in correspondence between a claim about the world and the way the world is.

On such a view of truth, for a statement to be true the world must be the way that the statement says it is. The statement must correspond to the facts. Thus, a theoretical claim about an unobservable entity is true if and only if the theoretical entity really is how it is claimed to be. For example, the sentence 'Electrons have negative charge' is true if and only if electrons have negative charge.[7]

It is important to note that there are a number of alternative theories of truth which contrast with the correspondence theory of truth. Among the most prominent of these are the coherence, pragmatist, consensus and internal realist theories of truth. According to such theories of truth, truth is a property which a belief or statement may have in virtue of some epistemic property of the belief or statement. Examples of epistemic properties which have been proposed by advocates of such theories of truth include internal coherence, practical utility, agreement with one's cultural peers, and ideal rational justification. Because such theories of truth identify truth with an epistemic property of belief, they are sometimes called 'epistemic theories of truth' (e.g., Devitt, 1991, 36, 44-5).

---

6 Hacking (1983, 93-4) suggests that no coherent sense may be made of the idea of a complete description of the world. Yet even if it were possible to make coherent sense of the idea, it seems unnecessary to burden the scientific realist with such a potentially objectionable assumption. The point is not that the realist may allow that there may be more than one complete true description of the world, but that the realist need not be committed to the possibility that there may be even one such description. (See 2.3 for further discussion.)

7 The issue of the truth of theoretical claims raises a question about theoretical discourse. Ian Hacking distinguishes between entity realism and theory realism (Hacking, 1983, 27). Entity realism asserts the reality of unobservable entities discovered by science. Theory realism asserts that scientific theories may be true or have a truth-value. Traditional scientific realism combines entity realism with theory realism. However, Hacking notes that the two doctrines are logically distinct. The entity realist may allow that there are unobservable entities of which scientists possess knowledge, but of which no current theory provides a correct description. By contrast, the theory realist may assert that a theory is true though none of its terms denote unobservable entities, but refer instead to logical constructions out of experience. In thesis 3, I have characterized the realist interpretation of theoretical discourse as a defining principle of scientific realism. Given this, it is not possible for scientific realism to deny that theoretical discourse purports to refer to real unobservable entities. However, it is no great departure from scientific realism to assert the reality of theoretical entities while denying theory realism. Entity realism may therefore be considered a special version of scientific realism.

For present purposes, I will understand the correspondence theory of truth in a broad sense. If a theory of truth holds that a statement is true just in case a given state of affairs obtains, then it will count as a correspondence theory. This means that minimalist theories which take truth to be exhausted by the equivalence scheme '"P" is true if and only if P' are just as much correspondence theories as are the more substantive attempts to identify the relation of correspondence with causal relations between language and reality.[8] What is characteristic of all such theories of truth is that truth is a relation of correspondence that obtains in virtue of the world in fact being the way that it is said to be.

Correspondence theories of truth contrast sharply with epistemic theories of truth, such as coherence or consensus theories, which identify truth with epistemic properties of beliefs. As we shall see later in this book, epistemic theories of truth imply an idealist covariance of belief and reality, and therefore cannot be reconciled with realism about a mind-independent reality (see 7.6 and 8.2). Correspondence theories which treat truth as a relation between language and reality are the only theories of truth compatible with realism.

The sixth core component of realism makes explicit the relationship between the two preceding components of realism, namely, that it is the objective world that renders our claims about the world true or false:

6. *Objectivity of truth:* theories or claims about the world are made true (or false) by the way things are in the mind-independent, objective reality investigated by science.

This thesis may seem redundant in light of the two preceding doctrines. However, I choose to state the thesis as a separate doctrine in order to explicitly rule out possible non-realist interpretations of the correspondence theory of truth. It is possible to incorporate the correspondence theory of truth within an idealist metaphysics. For example, the idealist may treat truth as a relation of correspondence between statements and states of affairs that are either solely constituted by mental states, or jointly constituted out of sensory input from the external world and the conceptual contribution of the human mind.[9]

Thus, what the realist wishes to say about truth is not merely that there is an objective reality and that truth is a correspondence relation. In addition, the realist wishes to say that truth consists in correspondence with objective reality. Thus, whether or not a statement is true has nothing to do with whether we happen to

8 While not all minimalist conceptions of truth may count as correspondence theories of truth in the broad sense at issue here, at least some do. Paul Horwich, for instance, argues that his own minimalist conception of truth is able to embrace 'the idea that each truth is made true by the existence of a corresponding fact' (1990, 112). For more substantive theories of truth, compare the attempt by Hartry Field and Michael Devitt to analyze the relation of correspondence as a function of a relation of reference between terms and their extension, where the latter is in turn to be analyzed by means of a causal theory of reference (Field, 1972; Devitt, 1991, 29).

9 In choosing to explicitly add the thesis of the objectivity of truth to that of correspondence truth, I follow the lead of Michael Devitt, who remarks that the correspondence theory 'is compatible with absolutely any metaphysics' (Devitt, 2002, 14).

believe that it is true. Rather, the truth of the statement is entirely determined by how things stand in the world, independently of us. This is a further reason why the realist is unable to countenance an epistemic conception of truth, since such a conception of truth removes the dependence of truth on an objective reality.

In sum, thesis 4, the thesis of metaphysical realism, says nothing explicit about truth. It only says something about the nature of reality investigated by science. Thesis 5, the thesis of correspondence truth, says nothing about reality. It only says something about the nature of truth. Thesis 6, the thesis of the objectivity of truth, is needed in order to make explicit the relation between the thesis of mind-independence and the correspondence theory of truth. It combines the two preceding theses into the realist doctrine that it is the mind-independent world that makes our claims about the world true in a correspondence sense. The three theses about reality, truth, and the relation between reality and truth are distinct theses. In the interest of clarity, they should be stated as such.

Optional Doctrines

I will now discuss a number of doctrines which are closely associated with, but not essential to, scientific realism. They are optional doctrines for realism. The first is the doctrine of semantic realism, which may seem to be indispensable to scientific realism. Then there is a cluster of related metaphysical doctrines which are endorsed by many scientific realists, but which are logically distinct from the scientific realist position that I have outlined in the preceding section.

The first optional component of scientific realism is a thesis about the meaning of empirical claims about the world. This is the semantic realist thesis that the meaning of the synthetic statements of empirical science consists in the conditions under which they would be true. According to semantic realism, the meaning of a scientific claim about the world consists in the conditions under which the claim would be true, rather than the conditions under which the assertion of such a claim would be warranted. In some cases, however, the truth-conditions of a scientific claim may be unable to be verified to obtain. Its truth-conditions may be 'verification-transcendent'.

In particular, it is impossible to conclusively establish by observational means alone the truth or falsity of theoretical claims about unobservable states of affairs (e.g., 'Electrons have negative charge'). Similarly, it is impossible to establish by such means the truth of universal generalizations about observable entities or states of affairs (e.g., 'All ravens are black'). Yet despite the inability to determine whether the truth-conditions of such claims obtain, the semantic realist holds that such claims may constitute significant assertions about the world, and indeed that they may possess a truth-value. This gives point to the attempt of scientists to provide indirect evidence for theoretical claims and empirical generalizations whose truth cannot be established by direct, empirical means.

Semantic realism contrasts with the strict verificationism of the early logical positivism of the Vienna Circle (e.g., Schlick, 1938). According to strict verificationism, the meaning of a synthetic claim consists in the empirical conditions under which it is verifiable, or in the means by which it may be empirically verified. On a strict construal of this doctrine, theoretical claims about the world are devoid of meaning or 'cognitive significance', because they are incapable of empirical verification. Later positivists reduced the demand for strict empirical verification to the weaker condition of non-conclusive confirmation. In so weakening the requirement of verification, the positivists conceded that claims about the world may possess verification-transcendent truth-conditions which are of relevance to the meaning of such claims. In allowing the relevance of such truth-conditions to meaning, the later positivists were conceding ground to semantic realism.

The rejection of strict verificationism in favour of semantic realism was a significant stage in the historical transition from logical positivism to scientific realism. Though strict verificationism is a thing of the past, the issue lives on in the form of the debate between realist theories of meaning which analyze the meaning of sentences in terms of truth-conditions and anti-realist theories of meaning which analyze sentence meaning in terms of assertability-conditions. The reason that I choose to treat semantic realism as an optional tenet of scientific realism is that I wish to leave it open that an assertability condition or use-based theory of meaning might turn out to be consistent with the basic principles of scientific realism.[10]

Turning to metaphysical doctrines, there are a number of interconnected realist views about causation and laws of nature. According to causal realism and nomological realism, causal and nomological relations are real relations of natural necessitation. Both causal and nomological realism stem from the rejection of the Humean empiricist view that causation and laws of nature are to be conceived as regularities. Where Hume held causal connections to be contingent conjunctions of events, causal realists treat causation as natural necessitation. When one event causes another to occur, it is not just that one event temporally precedes the other. Rather, the occurrence of the causing event necessitates the occurrence of the event which is its effect. Similarly, on a Humean treatment of laws of nature, the laws of nature are no more than contingent, empirical regularities. For the nomological realist, by contrast, genuine, non-accidental regularities are themselves the manifestation of real, underlying laws of nature. Events which occur as the result of a law of nature do so as the result of natural necessity.

A further optional doctrine is the doctrine of realism about natural kinds. According to natural kind realism, the theoretical entities of science are members of natural kinds. Some contemporary natural kind realists also espouse an essentialist view of kinds, according to which natural kinds are characterized by the possession of essential properties. It might at first seem that commitment to the reality of theoretical

10 Paul Horwich (1990, 72) argues that understanding a sentence involves knowledge of use conditions rather than truth-conditions. If this is right, then it would be possible to defend a use-condition theory of meaning, rather than a semantic realist account of meaning, while at the same time embracing a full-blown scientific realist position equipped with a correspondence theory of truth. The point is that meaning need not be understood in terms of truth-conditions in order to defend the view that science both aims for truth and generates knowledge of the truth. While I favour a semantic realist theory of meaning, the reason I do not wish to build semantic realism into the principles of scientific realism is that I see no reason to foreclose the possibility suggested by Horwich.

entities commits scientific realism to realism about natural kinds, since entities presumably belong to kinds. However, some scientific realists assert the existence of natural kinds, while others deny their existence. Indeed, it is possible both to assert the reality of theoretical entities and to deny realism about natural kinds. One may assert that electrons exist without asserting that the set of electrons forms a natural kind. The crux of the matter is the existence of kinds, not that of entities. One may assert that certain entities exist, but deny that they all have any essential property in virtue of which the entities belong to a kind. Alternatively, one may assert the reality of a group of entities which have many common features, without asserting the existence of a natural kind 'over and above' particular entities.

Neither causal realism, nomological realism, nor natural kind realism are essential doctrines of scientific realism. Endorsement of the core doctrines of scientific realism does not automatically commit one to views about such matters. However, all three doctrines form part of what one might broadly characterize as a 'realist outlook'. Causal and nomological realism fit with the realist rejection of the metaphysics lying behind traditional Humean empiricism. Natural kind realism reflects the realist intuition that the world is not something amorphous or devoid of structure, and that the structure of the world is there to be discovered. Moreover, nomological realism and natural kind realism also complement each other, since it is possible to argue that the necessity of laws of nature is in fact grounded in the essential properties of members of natural kinds (cf. Ellis, 2001).

1.3 Scientific Realism and Scientism

So far, I have sought to present the basic outlines of scientific realism. I would like now to further clarify scientific realism by contrasting it with a doctrine with which it is sometimes confused. Scientific realism is sometimes taken to be a form of *scientism*. Scientism is the doctrine that science is the sole legitimate source of empirical knowledge. The doctrine of scientism may be expressed in various ways, for example, as the positivist doctrine that cognitive significance is restricted to verification by means of scientific observation. A recent expression of scientism is the naturalist doctrine that the methods of natural science provide the sole means of epistemic access to the world. But while scientific realism may sit comfortably with the attitude of scientism, it would be a mistake to identify scientific realism with scientism.

First, let me note that it is no part of the family of scientific realist doctrines presented here that science is the only means of epistemic access to the world. The issue of knowledge only figures explicitly in the context of thesis 2, the thesis of epistemic realism, which asserts that science leads to knowledge of the objective world. But this in no way implies that science is the sole source of empirical knowledge. It is consistent with epistemic realism, as it is with the remaining principles of scientific realism, to allow that commonsense experience is a perfectly acceptable means of access to the world. Nor is there any reason for the scientific realist to hold that there may be no empirical knowledge in disciplines, such as history, which might fail to employ the methods of natural science. Thus, scientific realism does not entail a scientistic attitude toward knowledge. That is not to say, though, that it is incompatible with such an attitude. To see this, I will now briefly trace a route that leads from scientific realism to scientism.

Let us suppose that the only things that exist are material objects. Let us also suppose that all material objects are constituted out of the fundamental constituents of matter of which we are informed by physical science (atoms, molecules, and so on). Finally, let us suppose that anything which is not either a fundamental constituent of matter, or made up of a fundamental constituent of matter, does not really exist. This austere metaphysical picture is the metaphysics of materialism, or, to use a more contemporary name, physicalism. Such a materialist metaphysics is consistent with scientific realism, since it may be arrived at on the basis of realist commitment to the existence of theoretical entities. However, it is not entailed either by realism about theoretical entities, or by scientific realism in general, because physicalism is based on the further assumption that the only things that exist are the entities described by fundamental physics, and things which are made up out of such entities.

We are constrained in what we can know by the nature of what exists. Hence, ontology constrains epistemology. If physical things are all that exist, then this imposes certain constraints on our knowledge. We may only acquire knowledge of the properties of physical things by means that are available within a physical world. Since we are ourselves physical beings, our only means of epistemic access to the physical world must be by way of our causal interaction with it. Either we acquire knowledge directly by means of causal interaction of the world with our perceptual apparatus, or we acquire knowledge indirectly by means of causal relations with things to which we do not have direct perceptual access.

The way is now clear to scientism. For if we are prepared to assume that the methods of science represent the sole epistemic use of causal relations with the physical world, we may conclude that science is the sole means of epistemic access to the world. I do not myself find this an intrinsically abhorrent conclusion, as many do. But it should be clear that the route from scientific realism to such a scientistic conclusion is an indirect one, which requires further epistemological and metaphysical assumptions that play no part in the doctrine of scientific realism itself. It should also be clear that one may embrace scientific realism without taking the first step down the path to scientism.

1.4 Arguments for Scientific Realism

In the remainder of the chapter, I will present what I take to be the main arguments in favour of scientific realism. I noted before that scientific realism does not consist in any one, single doctrine. It is an interconnected family of theses. Similarly, there is no one, single argument for scientific realism. The so-called 'success argument' has received a great deal of attention. But there are other arguments which deserve equal billing. There is no doubt that the success argument is an important part of the case for scientific realism. But it only comes into play once the ground has been laid by other realist lines of argument.

In what follows, I will not mention all known arguments for scientific realism, but only those which form part of the overall case that I will present for scientific realism.[11] The first two lines of argument turn on reflection upon our place in the natural order and an appeal to the epistemology and metaphysics of common sense. The second two arguments are versions of the success argument. The first version relates to the truth or approximate truth of theories, while the second version of the argument applies success at the level of the methods employed in science.

Our Place in Nature

The basic argument for realism takes off from a founding intuition about our place in nature. We human beings are sentient, intelligent organisms. We inhabit a pre-existing natural world. We interact causally with this world. But we did not invent, create or construct it. We must act in the world in order to survive. To assure our survival, we must acquire knowledge of the way the world is. For knowledge about the way the world is enables us to reliably undertake actions which promote our survival. Thus, the realist concludes, we are creatures who inhabit an objective reality, of which, given our survival, we have the capacity to acquire genuine knowledge.

This is the perspective of realism, spiced, I should say, with a dash of evolutionary naturalism. The perspective is fundamentally opposed to views which conceive reality on the basis of human mental representation, such as belief or experience. The realist sees humans and their inner life as but a small part of a vast reality. Any view which takes human thought or experience as the basis of reality, or of the concept of reality, profoundly misunderstands our place in the natural world. From the realist perspective, such a view commits the fundamental error of anthropocentrism (cf. Smart, 1963; also Hooker, 1987, 264ff).

The realist takes the external world as a given. The existence of the external world does not depend on thought or experience. It is a world in which we find ourselves embedded and which we inhabit. We are able to effect change in the world by means of actions which bring about such change. We construct buildings, grow crops, and pollute the environment. But we did not make the world. Nor do the basic entities which populate the world, or the laws of nature which govern the behaviour of these entities, depend on us in any way. Thus, rather than take human thought or experience as primary in forming our conception of reality, the realist takes human thought and experience to form a part – indeed, a relatively insignificant part – of that reality.

The opposing non-realist perspective has its origins in the sceptical problematic of traditional epistemology. Scepticism of the Cartesian variety challenges us to show that there is an external world and that we have knowledge of such a world. Traditional epistemologists took mental representations as the basis of their response

11 Putnam (1975a, 72) distinguishes between negative and positive arguments for realism. I will ignore negative arguments for realism, though these have been historically very powerful arguments for realism. Negative arguments are arguments against opposing positions, examples of which include the series of arguments proposed in the late 1950s against the logical empiricist treatment of theoretical discourse.

to such scepticism. On the basis of beliefs, ideas or experience, they sought to show both that there is an external world and that we are able to have knowledge of it. Philosophers who attempt to meet the sceptic in this way typically find that the game is rigged against them. For the sceptic sets the standards too high, demanding absolute certainty where none is to be found.

As against traditional, sceptic-centered epistemology, the realist takes it as a basic starting-point that there is an external reality, and, indeed, that we are able to have knowledge of that reality. For the realist, the lesson of scepticism is not that knowledge of the external world is impossible, but that it is a mistake to seek epistemic certainty or to treat mental representations as the basis of either our epistemology or metaphysics. We know just as surely as we may know anything that there is an objective, external reality, and that we may come to have knowledge of it. But our knowledge need be neither certain, nor grounded in privileged representations of that reality.

Thus, from the perspective of realism, it is a mistake to base our concept of reality on human mental representation. Mental representations are but a small part of a greater reality in which we find ourselves embedded. Any philosophy which seeks to ground our conception of reality on our own mental representations commits the fundamental error of anthropocentrism, and should therefore be dismissed as fatally flawed.

Commonsense Realism

The second strand in the argument for scientific realism turns on an appeal to common sense and the realism implicit in ordinary common sense. By 'common sense', I mean our ordinary, prereflective awareness of our immediate surroundings and of the broader world which extends beyond those immediate surroundings. This is a world that is made up of material objects of all shapes and sizes, of which we have more or less immediate knowledge by means of our sensory experience of those objects. It is a concrete world of mind-independent objects with which we interact causally by means of bodily movement and action, but which is nonetheless beyond the immediate control of our powers of volition. It is also a world in which misperception and illusion have their place in the ordinary course of events, but in which a robust sense of reality nevertheless sustains a reasonable degree of practical certainty that things are by and large as they seem.

Realism about ordinary everyday objects and our epistemic access to such objects provides the starting-point for the commonsense realist component of the argument for scientific realism. Common sense gives rise to a body of beliefs about the objects in our environment and our epistemic and practical interactions with these objects. On the whole, we may assume that this body of beliefs is true. The point is not that our commonsense beliefs are certain, indubitable or infallible. Rather, commonsense beliefs are *prima facie* justified. They have an epistemic priority, which makes them difficult to dislodge by rational argument. Any attempt to eliminate or overthrow such beliefs is to be regarded with extreme suspicion. Any argument that purports to show that common sense is to be discarded thereby shows itself more than likely to be unsound or invalid.[12]

Such a robust, commonsense attitude underwrites commonsense realism about ordinary, everyday material objects and our perceptual access to such objects. The scientific realist who takes common sense as a starting-point is thereby justified in assuming that there is an ordinary, everyday world of material objects, with which we interact causally and to which we have epistemic access by means of our senses. The scientific realist is free to build upon the basis of commonsense realism in arguing that scientific theories, realistically construed, are the best explanation of observed phenomena at the commonsense level. There is no need for the scientific realist to argue for the reality of ordinary, everyday material objects, since commitment to such entities has already been established at the level of common sense.

While the attitude of common sense leads to realism about the objects of common sense, such realism contains the seeds of a more full-blown realism about scientific theories and entities. For one thing, a tendency toward realism about scientific theories and entities is built into commonsense realism about ordinary objects. While we may be unable to observe the basic constituents of material objects with our naked eyes, we are accustomed to the idea that material objects have component parts, and that some of these parts may be too small to see. The full-blown scientific view that matter is composed of fundamental particles, atoms and molecules is but a highly sophisticated extension of the commonsense idea of the compositional nature of matter.

For another thing, commonsense realism treats the objects of ordinary common sense as real, objective entities, which exist independently of human mental activity. Scientists from different historical epochs, or scientists who work in different Kuhnian 'paradigms', occupy the same commonsense world of ordinary, everyday objects. Because scientists from different historical epochs inhabit the same commonsense world, modern scientists confront the same observable objects and phenomena as did ancient scientists who worked in the same domain. Equally, proponents of alternative Kuhnian 'paradigms' do not inhabit different 'worlds', but maintain common perceptual access to a shared domain of observable objects (cf. Kuhn 1970a, 111, 150).[13]

While common sense coheres well with scientific realism, it must be admitted that a certain tension may sometimes arise between science and common sense. Here is a familiar example from the history of astronomy. Our senses tell us that the Earth is flat. Yet science tells us that the Earth is spherical. Our senses tell us that the sun moves across the sky each day, rising in the East and setting in the West. Science tells us that it is the daily rotation of the Earth that makes the sun appear to move. Our senses tell us that the Earth is immobile. Yet science tells us not only that the Earth rotates upon its axis, but that it revolves around the sun in an annual orbit.

12 The point that common sense is more likely to be correct than any philosophical argument against it is emphasized by Armstrong (1999), Campbell (1988) and Devitt (2001), who credit the basic thought to G.E. Moore.

13 The point that common sense tells against the incommensurability of paradigms is well made by Campbell (1988).

Such apparent conflicts between science and sensory evidence have led some realistically inclined philosophers to hold that there is an inherent tension between science and common sense (Sellars, 1963; Feyerabend, 1975; Churchland, 1979). Common sense is the repository of primitive theory. It is the 'metaphysics of the stone age', in Russell's words. With the advance of science, such primitive theory is inevitably corrected, refuted and ultimately eliminated. Thus, by the lights of science, common sense must itself be rejected. Common sense cannot therefore serve as the basis for a realist account of science.

It is undeniable that conflict may on occasion arise between science and common sense. But it is an exaggeration to inflate such conflict into a fundamental incompatibility between science and common sense. In such conflict, the commonsense description of the phenomenon is typically not corrected by science at all. What science corrects is the *explanation* of the appearances.

The Earth *appears* flat. The sun *appears* to move across the sky each day. Science places the appearances within the context of a theoretical system, which corrects the commonsense view by explaining how the rotation of a spherical Earth gives rise to the appearance of the sun's daily transit across the heavens. This is precisely a case in which common sense is only renounced in favour of an improved explanatory structure, which both preserves and explains the appearances noted by commonsense observation.

Success and Truth

As we have just seen, commonsense realism contains the seeds of scientific realism. There is a further sense in which this is the case. In the course of everyday practical activity, we routinely employ inference to the best explanation in seeking to understand why various events occur. Such reasoning is the basis of the best-known argument for scientific realism, the so-called success or 'no miracles' argument.[14] The reasoning that forms the basis of one of the major arguments for scientific realism is therefore reasoning of a commonsense kind.

The classic formulation of the success or 'no miracles' argument is due to Hilary Putnam:

> The positive argument for realism is that it is the only philosophy that doesn't make the success of science a miracle. That terms in mature scientific theories typically refer (this formulation is due to Richard Boyd), that the theories accepted in a mature science are typically approximately true, that the same term can refer to the same thing even when it occurs in different theories – these statements are viewed by the scientific realist not as necessary truths but as part of the only scientific explanation of the success of science,

---

14 In addition to the success argument, there are a number of other positive arguments for scientific realism, e.g., Wes Salmon's argument from the common cause (Salmon, 1984, 206ff) and Ian Hacking's direct 'experimental proof' of realism (Hacking, 1983, 265). But such arguments may be assimilated to the success argument. For example, the existence of an entity which is the common cause of a number of different phenomena is the best explanation of those phenomena. Similarly, the existence of an unobservable entity which produces certain experimental results is the best explanation of successful laboratory practice.

and hence as part of any adequate scientific description of science and its relations to its objects. (Putnam, 1975a, 73)

In this passage, Putnam argues that realism is the best explanation of the success of science. (Strictly, he says it is the *only* explanation, but this is a form of inference to the best explanation.) Putnam's argument turns on the claim that a philosophy of science which denies that theoretical entities are real, or that scientific theories are true or approximately true, must treat the success of science as a miracle that is incapable of explanation. An explanation which treats the success of science as an inexplicable miracle is an unsatisfactory explanation of such success. By contrast, scientific realism provides a compelling explanation of the success of science. On the whole, the unobservable entities postulated by theories exist, and scientific theories are true or approximately true. Given the reality of the entities to which scientific theories refer, as well as the truth or approximate truth of such theories, it is only to be expected that science should manifest the striking degree of empirical success that it does. Because scientific realism provides a compelling explanation of the success of science, while alternative approaches provide an unsatisfactory explanation, we should accept scientific realism as true.[15]

Various objections have been raised against the success argument. Of particular relevance in the present context are historical counterexamples to the success argument due to Larry Laudan (1984, Ch. 5). Laudan presents a list of historical cases of scientific theories (e.g., eighteenth century chemical atomism, Wegener's continental drift theory), now considered to have been approximately true or referential, but which met with little or no success in their time. He also presents cases of successful theories (e.g., the ether and phlogiston theories) which are now thought neither to have been referential nor to have been true or approximately true. Laudan's counterexamples appear to show that there is no connection between the empirical success of a theory and reference, truth or approximate truth. If he is right, the claim that scientific realism is the best explanation of the success of science would appear unsustainable.

15 In my gloss of Putnam's 'no miracles' argument, I have also drawn upon the following passage from Putnam (1978), which speaks less about reference and more about the entities referred to by theories:

... the modern positivist has to leave it without explanation (the realist charges) that 'electron calculi' and 'space-time calculi' and 'DNA calculi' correctly predict observable phenomena if, in reality, there are no electrons, no curved space-time, and no DNA molecules. If there are such things, then a natural explanation of the success of these theories is that they are *partially true accounts* of how they behave. And a natural account of the way in which scientific theories succeed each other – say, the way in which Einstein's Relativity succeeded Newton's Universal Gravitation – is that a partially correct/partially incorrect account of a theoretical object – say, the gravitational field, or the metric structure of space-time, or both – is replaced by a *better* account of the same object or objects. But if these objects don't really exist at all, then it is a *miracle* that a theory which speaks of gravitational action at a distance successfully predicts phenomena; it is a *miracle* that a theory which speaks of curved space-time successfully predicts phenomena... (Putnam, 1978, 19)

Recent work by Kitcher, Musgrave and Psillos suggests that the success argument may be revised in a way that renders it immune to Laudan's criticism. For one thing, if the criterion of scientific success is revised to include only those theories which exhibit a high degree of novel predictive success, then a number of Laudan's counterexamples may be dismissed as not displaying the requisite degree of success.[16] For another thing, if credit for success is restricted to the constituents of a theory which are responsible for novel predictive success, this increases the likelihood that the relevant constituents will be preserved in the course of subsequent theory modification, and later considered approximations to the truth.[17]

I regard such revisions of the success argument as well-motivated. In attempting to determine whether a successful theory is true, it is important to employ a rigorous standard of success such as novel predictive success. It is also important to assign credit for such success to the constituents of theory specifically responsible for such success. But the result of so revising the success argument is a weakened position which fails to meet the epistemological needs of scientific realism. The realist does not merely wish to defend a claim about the truth of theories. It is of at least equal importance for the scientific realist to defend the epistemic realist view that the methods of science produce rationally justified belief, and indeed knowledge, with respect to those aspects of the world about which scientific theories purport to inform us. Hence, the scientific realist must also defend a realist epistemology for science.

Success and Method

In the appraisal of a scientific theory, and the choice between alternative theories, scientists employ a variety of rules of method. They consider whether a theory is confirmed by the evidence, accurately predicts novel facts, unifies phenomena from disparate domains, and so forth. If a theory is certified by such rules of method, then a scientist is rationally justified in accepting the theory. Certification by rules of method therefore provides the basis for epistemic warrant in science.

The scientific realist wishes to defend the epistemic realist thesis that scientific inquiry leads to rational belief and knowledge about unobservable aspects of the world. The realist must therefore argue that use of the rules of method gives rise to theories which scientists are warranted in accepting as true or approximately true. For this reason, while I am favourable to the revisions of the success argument noted above, I suggest that emphasis should be placed instead on application of the success argument at the level of the methods of science.[18]

---

16 The importance of novel predictive success is urged by Musgrave (1999, 55) and Psillos (1999, 105).

17 For the claim that credit for the success of a theory should be accorded to only those parts of a theory responsible for the success, see Kitcher (1993, 143-9) and Psillos (1999, 108). For critical discussion of this claim, see Lyons (2006) and Stanford (2002).

18 The strategy of applying the success argument at the level of method has been championed by Richard Boyd (e.g., 1984, 58ff). In related ways, it has also been employed by Rescher (1977) and Kornblith (1993a). In my attempt to develop a naturalized epistemology for scientific realism, it is the strategy that I employ as well (see Chapters 6-8).

In particular, I propose an approach that I call *abductive realism* (see 7.9). According to this approach, the best explanation of the cognitive and pragmatic success of scientific theory and practice is that the rules of method are truth-conducive tools of inquiry, which serve as reliable means for obtaining truth.

Abductive realism forms part of a naturalistic theory of epistemic warrant. This theory treats rules of method as cognitive instruments, which serve as means for the achievement of epistemic ends. Such an instrumental construal of the rules of method enables the question of the warrant supplied by a rule to be understood as the empirical question whether use of the rule conduces to the epistemic end it is claimed to promote. Rules of method which reliably promote the aim of truth provide scientists with epistemic warrant for accepting theories which satisfy those rules. Thus, the normative force of rules of method is grounded in empirical facts about effective means of inquiry into the mind-independent, natural world which we inhabit.[19]

Abductive realism addresses the question of why the rules of method are to be taken to promote the realist aim of truth. I call it an *abductive* strategy because it is based on inference to the best explanation, a form of abductive inference. An inference of this kind is required because of the lack of direct evidence for the connection between method and the truth of theory. Because the truth of the non-observational content of theories cannot be established by observation, no connection between method and the truth of theories may be shown to obtain by empirical means. Thus, the grounds for taking the rules of method to be truth-conducive can be at best abductive grounds.

Abductive realism places special emphasis on the regulative role of method in the selection and elimination of theories. The rules of method serve as a means of 'quality control'. Scientists employ rules of method as selection criteria on the basis of which to eliminate faulty theories in favour of ones that are serious contenders for truth.

The regulative role of the rules of method enables them to serve as the arbiter of success. Suppose that a theory satisfies the rules of method to a remarkably high degree. It accommodates all known data, and accurately predicts many surprising novel facts. It unifies disparate domains in a simple and coherent manner, while opening up exciting new areas of inquiry. From a methodological point of view, such a theory is an ideal theory.[20] It manifests a near perfect level of success. According

19 I develop this general line of argument in Chapters 6 and 7. For the naturalistic view that the rules of method may be viewed instrumentally as means to cognitive ends, see, Laudan (1996), Rescher (1977) and Stich (1990).

20 I do not assume that the methodologically ideal theory is the theory that will be reached at the ultimate end of ideal inquiry. Rather, it is an ideal theory which might be reached at some more mundane point of inquiry. However, it is worth briefly addressing the issue of the methodologically ideal theory reached at the ultimate end of ideal inquiry. Because the scientific realist takes truth to be defined as correspondence between language and reality, rather than in terms of ideal satisfaction of epistemic criteria, the theory reached at the ultimate end of ideal inquiry could well be false, as Putnam suggests is the case for the doctrine he calls 'metaphysical realism' (1978, 125). While Putnam's internal realism identifies truth with ideal rational justification, scientific realism in the form that I have presented the position

to abductive realism, the best explanation of such success is that the rules of method are regulative norms which 'screen for truth'. They are genuinely truth-conducive instruments of inquiry, which rigorously select only those theories which are either true, or on the track of truth.

It does not suffice, of course, to simply assert that realism is the best explanation of ideal satisfaction of method. An argument is needed. Here the abductive realist employs a metamethodological analogue of the classic 'no miracles' version of the success argument. How might ideal methodological success be explained by the opponent of realism? Let us focus on the straightforward opponent of scientific realism. Such an anti-realist denies the realist's claims about truth and reference. The ideal theory is neither true nor approximately true. Its terms fail to refer to any real thing. None of the entities postulated by the theory exist.

Such an anti-realist is entirely without the resources to explain ideal methodological success. If a theory fails not only to be true but even approximately true, and none of its terms refer to any real entities, then the success of such theory is nothing short of a miracle. But that is surely not an explanation of the success of science.

1.5 Conclusion

I will briefly conclude by commenting on some of the relations between the arguments I have offered for scientific realism and the various doctrines which comprise the position of scientific realism. Realism about the external world (thesis 4) is supported by the rejection of anthropocentricism and the appeal to common sense. Because science is an extension of common sense, a realist treatment of theoretical discourse (thesis 3) derives general support from common sense, though it derives more direct support from the appeal to the success of science. The epistemic realist thesis (thesis 2) that we have genuine knowledge of unobservable aspects of reality gains broad support from common sense. However, it is supported more directly by the success argument, and most directly by the metamethodological application of the success argument.

This does not exhaust the connections between arguments for realism and various principles of realism. But the fact that different strands of realist argument bear on different components of realism further illustrates my main point. Scientific realism is not captured by any one doctrine. It is a complex position. Because it is a complex position, different lines of argument must be brought to bear in support of different aspects of the position.

---

here opposes any such identification. But denial of such identification does not debar the scientific realist from holding that the best explanation of ideal methodological success is that the theory reached at the end of ideal inquiry is true in the realist correspondence sense.

Chapter 2

The God's Eye Point of View

2.1 Introduction

Hilary Putnam once claimed that the position he referred to as metaphysical realism presupposes a God's Eye point of view. That is, it presupposes that we are able to remove ourselves from our human perspective and survey the world as it really is from the point of view of an omniscient being. But it is impossible to remove ourselves from our human perspective in the way that metaphysical realism requires. So, Putnam argues, metaphysical realism is an unsustainable position.

With some minor qualifications, which I will note below, I take scientific realism to be a form of metaphysical realism. Hence, I will tend to use the terms 'realism' and 'scientific realism' interchangeably. According to scientific realism, as I understand it, scientific inquiry leads to knowledge of the truth about observable and unobservable aspects of a mind-independent, objective reality. The scientific realist's commitment to a mind-independent reality qualifies scientific realism as a form of metaphysical realism. Because scientific realism is a form of metaphysical realism, the problem of the God's Eye view must also arise as a problem for the scientific realist.

My aim in this chapter is to defend scientific realism against the Putnam-style objection that it incoherently requires a God's Eye point of view. In particular, I seek to establish two points. First, scientific realism does not (incoherently) presuppose a God's Eye point of view. Second, even if scientific realism did presuppose a God's Eye point of view, this would provide no basis on which to object to scientific realism.

2.2 Putnam on the God's Eye Point of View

The idea that realism requires a God's Eye point of view is found in the third chapter of *Reason, Truth and History*, where Putnam rejects metaphysical realism in favour of internal realism.[1] Putnam characterizes metaphysical realism in the following terms:

---

1 As it is now more than 20 years since the publication of *Reason, Truth and History*, and Putnam has long since moved on from the internal realism he then advocated, the topic of the God's Eye point of view may seem somewhat dated. However, the problem of the God's Eye point of view has caught hold amongst some philosophers. Two recent authors, Marsonet (2002) and Tetens (2004), write as if the problem of the God's Eye point of view exposes a serious shortcoming in the realist position. It is because I disagree with authors such as Marsonet and Tetens on this score that I have been prompted to offer my response to the

> On this perspective [i.e., the perspective of metaphysical realism], the world consists of some fixed totality of mind-independent objects. There is exactly one true and complete description of 'the way the world is'. Truth involves some sort of correspondence relation between words or thought-signs and external things and sets of things. I shall call this perspective the *externalist* perspective, because its favorite point of view is a God's Eye point of view. (Putnam, 1981, 49)

Following this passage, Putnam goes on to characterize the internal realist perspective that he favoured at that stage in his philosophical development. He then comments that:

> There is no God's Eye point of view that we can know or usefully imagine; there are only the various points of view of actual persons reflecting various interests and purposes that their descriptions and theories subserve. (Putnam, 1981, 50)

Given that there is no God's Eye view available to us, metaphysical realism is not a position that can be justifiably adopted. For to adopt metaphysical realism would require us to occupy the standpoint of God, which is impossible for us.

The internal realist position that Putnam proposes in *Reason, Truth and History* involves a conception of truth that reflects Putnam's rejection of the God's Eye perspective. The central tenet of internal realism is that truth is an 'idealization of rational acceptability' (1981, 55). Truth is what would arise if scientific inquiry were pursued to the ideal limit of inquiry.[2] It is 'some sort of ideal coherence of our beliefs with each other and with our experiences *as those experiences are themselves represented in our belief system*' (1981, 49-50).

The internal realist conception of truth is therefore an epistemic conception of truth. As such, it contrasts with the metaphysical realist view of truth, according to which truth is a non-epistemic relation of correspondence between language and reality (1981, 55). The internalist's epistemic conception of truth is therefore a repudiation of the God's Eye point of view. For the internalist denies that truth may be conceived as independent of either observer or conceptual scheme, as required by the metaphysical realist's God's Eye view of truth.

In sum, I take the thrust of Putnam's comments about the God's Eye point of view to be twofold. On the one hand, realism requires a God's Eye point of view in order to be stated or defended. On the other hand, it is impossible for us to adopt a God's Eye point of view. So realism is incoherent. For it is not possible for us to occupy the standpoint that we would need to occupy in order to be able to formulate or to defend the position.

---

problem in this chapter. I specifically discuss the views of Marsonet in the Appendix to this chapter. For my comments on Tetens, see Sankey (2004).

2 Cf. Putnam (1978, 125), where Putnam notes that metaphysical realism treats truth as 'radically non-epistemic', which implies that the ideal theory reached at the ultimate end of scientific inquiry might be false. Though Putnam does not, in so many words, assert that the ideal theory is true, this is the clear implication of his internal realist identification of truth with ideal rational justification.

2.3 Metaphysical Realism and Scientific Realism

The specific target of Putnam's God's Eye objection is the position of metaphysical realism rather than scientific realism, as such. Still, I understand scientific realism to be a form of metaphysical realism. Thus, while I have a number of reservations about Putnam's characterization of metaphysical realism, I take the God's Eye objection to apply to scientific realism as a special case of metaphysical realism. Before considering whether scientific realism requires a God's Eye view, let me briefly indicate the sort of qualifications that I believe need to be made about Putnam's characterization of metaphysical realism.

In the first place, it is not clear that the realist need be committed to the existence of 'one true and complete description of "the way the world is"', as Putnam suggests in the passage quoted above. To avoid relativism about truth or reality, it must of course be denied that there is more than one true and complete description of the world.[3] But the core commitment of metaphysical realism is to the existence of an objective reality whose existence, properties and structure are independent of human mental activity. It is an open question whether there need be even as many as one true and complete description of such a reality. Indeed, it is an open question whether coherent sense may be made of the idea of such a complete description.[4]

In the second place, it is not clear that the realist need be committed to the view that 'the world consists of some fixed totality of mind-independent objects'. For, as Alan Musgrave has pointed out in discussion of Putnam, the word 'object' is not an individuating or sortal expression (2001, 41). The question of how many objects exist is not, therefore, a well-formed question. Such a question has no answer unless a specification is given of what kind of object one has in mind. But this means that the claim that the world consists of a *fixed totality* of objects is not one to which any clear significance may be attached. While the realist might be committed to the existence of a fixed totality of some *specific kind(s)* of objects, there is no need – nor does it make any sense – for the realist to be committed to a general claim that there is a fixed totality of objects.

But let me set such reservations about Putnam's characterization of metaphysical realism to one side. In the previous chapter, I have characterized scientific realism in terms of six core principles (see Section 1.2). This characterization of scientific realism departs from Putnam's metaphysical realism in a number of key respects. But I do not think that it does so in a way that would render it immune to Putnam's worry about the God's Eye view. For on my characterization of scientific realism, realism is committed to a non-epistemic conception of truth as correspondence to an objective, mind-independent reality. Such a conception of truth is surely the key constitutive component of metaphysical realism in the sense defined by Putnam.

3 More precisely, to avoid relativism, it must be denied that there may be true and complete descriptions of the world which are jointly inconsistent with each other. In principle, it might be possible to formulate alternative true and complete descriptions on the basis of alternative conceptual schemes. Provided that such descriptions are consistent with each other, no threat of relativism arises. (I owe this point to Michael Devitt.)

4 For sustained criticism of the idea of a complete description of the 'way the world is', see Hacking (1983, 93-5).

Thus, apart from the minor qualifications mentioned above, scientific realism in my sense is similar in spirit to metaphysical realism in Putnam's sense.[5]

2.4 Does Scientific Realism Require a God's Eye Point of View?

Let us now consider whether scientific realism requires a God's Eye point of view.

It may immediately be conceded that it is impossible for us to remove ourselves from our human perspective and adopt a God's Eye point of view. We are unable to survey the world from the vantage point of an all-knowing supreme being. But neither does the doctrine of scientific realism require that we adopt such a viewpoint.

The realist who proposes a scientific realist interpretation of science does not thereby purport to occupy a God's Eye perspective. Rather, in proposing such an interpretation of science the scientific realist puts forward a hypothesis about the nature of science and the relation between science and reality. In particular, the realist claims that science is an activity, the aim of which is to discover the truth about observable and unobservable dimensions of a mind-independent, objective reality. But such a claim is not made from a God's Eye point of view. It is a hypothesis that the realist proposes from within our human perspective as an interpretation of a specific human activity, the activity we call science.

Now, it would be perfectly consistent for the scientific realist to refrain from any positive epistemic commitment to the truth or progressiveness of science. A sceptically minded realist might adopt a restricted position about the aim of science and the interpretation of theoretical discourse, but suspend judgement on the question of whether any actual progress has been made toward the scientific aim of truth. In effect, such a restricted version of scientific realism would amount to suspending the principle of epistemic realism (thesis 2) while asserting the remaining five principles of scientific realism.

But realists typically do not adopt such a sceptical attitude toward science. They typically support a stronger epistemic thesis to the effect that science has made progress toward the truth, and, in so doing, has produced genuine knowledge about the objective world. Realists typically combine the realist interpretation of science with the additional claim that a realist interpretation of science provides the best explanation of the success of science. In particular, realists typically assert that the truth or approximate truth of scientific theories, together with the successful reference of theoretical terms, is responsible for the much vaunted empirical success of the sciences. Given this, realists conclude, the hypothesis of scientific realism should be accepted as an accurate portrayal of the relation between science and reality.[6]

---

5 In his (2005), Michel Ghins has objected to the metaphysical realist component of my scientific realist position by claiming that the mind-independent world 'lie[s] beyond our cognitive and linguistic abilities' (2005, 145). But the fact that the world is independent of cognition or language does not entail that we are unable to think or talk about it. We are able to think and talk about things that do not depend on thought or talk.

6 What I have just described, of course, is the well-known success argument for scientific realism. As we saw in Chapter 1, the success argument is only one part of the case to be made for scientific realism.

But such a hypothesis about the relation between science and reality makes no evident use of a God's Eye point of view. Quite the contrary, it is a hypothesis proposed from within our human perspective about the relation between science and reality. So, far from laying any claim to omniscience, or direct access to reality, the realist claims that scientific realism provides the best explanation of a robust phenomenon that stands in need of explanation, namely, the empirical success of the sciences. Far from presupposing a God's Eye perspective, the argument is designed to persuade fellow occupants of our human perspective that a realist account of science provides the best account of the epistemic and semantic relations between the human activity of science and the largely non-human world that we inhabit.

It is worth remarking that the strategy I have just sketched of arguing that realism is a hypothesis which is to be accepted because it provides the best explanation of the success of science reflects a broadly naturalistic conception of realism in the philosophy of science. For to treat realism as a hypothesis about science, and to argue for the hypothesis of realism by means of inference to best explanation, is to treat realism as a hypothesis that is to be evaluated in a manner analogous to the evaluation of scientific hypotheses. As such, my claim that realism does not require a God's Eye point of view derives from a naturalistic attitude toward the position of realism.[7]

Of course, it might be objected that realism is not the best explanation of the success of science. Instead, it might be maintained that there is a non-realist interpretation of science that provides a better explanation of the success of science than realism does. Or else it might simply be held that realism is to be rejected as an inadequate explanation, since truth and reference are not the invariable correlate of scientific success.

But such objections would be beside the point. For the point at issue is not whether realism provides the best explanation of the success of science, or, indeed, whether realism is true. Rather, the point at issue is whether the realist must adopt a God's Eye point of view in order to propose the realist hypothesis about the relation between science and reality. It might very well be the case that realism fails to be the best explanation, or is in fact an unsatisfactory explanation, of the success of science. But neither point has any bearing upon the issue of whether the realist must adopt a God's Eye point of view in order to propose a realist interpretation of science.

2.5 Is there a Coherent God's Eye Point of View?

So far, I have sought to show that scientific realism does not require a God's Eye point of view. As such, it does not fall prey to Putnam's objection that realism incoherently assumes a God's Eye point of view. I now wish to change tack and argue in the opposite direction. I will argue that, even if realism did appear to require

7 A related treatment of the issue may be found in Devitt (1991, Section 12.6). I am grateful to Michael Devitt for drawing my attention to this point, and for prompting me to explicitly note the naturalistic provenance of my argument in this section. As will be seen in the next section, this same naturalistic approach may also be used to defuse the threat of an appeal to the God's Eye point of view.

the adoption of a God's Eye point of view, this would not necessarily be to the detriment of realism. For coherent sense may be made of an external viewpoint suitable for realist purposes, which does not require us to adopt the perspective of an infallible, omniscient deity. I shall approach the issue from the perspective of a naturalized epistemology set within a realist framework. My point of departure will be the question of animal knowledge.

Recent work in cognitive ethology on the nature of animal minds is increasingly attracting attention amongst philosophers. Much of the interest revolves around questions in the philosophy of mind of whether non-human animals have minds in anything like the sense in which humans have minds. Considerable attention has also been focused on the ethical dimensions of our relations to other animals, since at least some non-human animals may be capable of a mental life in ways that have moral significance.

But the question of most immediate relevance relates to the epistemological significance of animal minds. Assuming that some animals may have minds, could the mental states of animals be epistemic states? Can animals other than us have knowledge?

In his recent book, *Knowledge and its Place in Nature* (2002), Hilary Kornblith presents a thoroughly naturalistic account of knowledge. Rather than analyze the concept of knowledge in the manner of traditional analytic epistemology, Kornblith proposes that epistemologists should treat knowledge as a natural kind which may be investigated by empirical means. But, if knowledge is a naturally occurring phenomenon, as this suggests, then the possibility arises that the natural kind, knowledge, may be instantiated in beings other than humans. It therefore becomes an open question, subject to empirical investigation, whether animals other than us are capable of knowledge, and what the nature of such knowledge might be.

Naturalistic philosophers, such as Kornblith, are often inclined to see continuities between humans and other animals. This raises a host of challenging philosophical questions. But, rather than explore any of these questions, I wish to focus on the implications of naturalism for the God's Eye objection. In particular, I wish to suggest that for philosophers who work from a naturalistic perspective, the problem of the God's Eye point of view poses less of a threat than might at first appear to be the case.

I will illustrate the point by means of one of the examples that Kornblith presents in *Knowledge and its Place in Nature*. Kornblith discusses the work of Carolyn Ristau on the piping plover (see Ristau, 1991). The piping plover is a shorebird found on the coast of eastern North America. Like a number of other plovers, such as the Killdeer, it employs deceptive, 'injury-feigning', behaviour in order to protect its young. When a human or other potential threat or predator approaches the nest, the adult plover attracts the intruder's attention by pretending to have a broken wing and moving away from the nest. Once the intruder has been led well away from the young, the plover flies off, leaving the intruder at some distance from the nest where it poses no immediate threat to the young birds.

As Kornblith points out, Ristau employs an epistemic idiom to describe the behaviour of the plover (Kornblith 2002, 53). The plover has knowledge of its environment. It knows whether an intruder poses a threat and it can determine whether an intruder is looking in the direction of the nest. It can discriminate between a person who has posed no threat in the past and one who has previously behaved in a dangerous manner. As the plover leads the intruder away from its nest, it continues to track the movements and position of the intruder, as well as to be aware of the location of its young, to ensure that the intruder is led away from the young.

All of this suggests that piping plover distraction behaviour involves epistemic states on the basis of which the plover is able to behave in such a way as to lead intruders away from its young. Kornblith notes that Ristau does not herself argue for the attribution of knowledge, rather than true belief, though she defends attribution of intentional states at some length (2002, 55).[8] But it is not essential for present purposes to establish the legitimacy of knowledge attribution to the plover, as opposed to some weaker epistemic state. Instead, it suffices to reflect upon the standpoint of the researcher who conducts a study of the plover's behaviour, and who proposes an explanation of this behaviour in terms of mental states of the plover and their relation to its environment.

Such reflection will return us to the question of whether it may be possible to coherently adopt a God's Eye point of view. For I wish to suggest that work such as Ristau's on the piping plover illustrates how one might adopt an external point of view that is not dissimilar to the God's Eye point of view that Putnam claims to be beyond reach. To see this, let us consider what of philosophical significance might emerge from Ristau's studies of the piping plover's mental states and their relation to its environment. Such empirical studies will ultimately provide the basis for an epistemological analysis of the cognitive states and associated behaviour of the piping plover.[9]

The results of such an epistemological analysis will take the form of claims about how the plover acquires knowledge of its environment, integrates new knowledge with prior knowledge, and utilizes such knowledge as the basis for action. For example, the plover may detect and monitor the approach of an intruder using its eyes. On the basis of such input, the plover determines an appropriate trajectory by which it can lead the intruder away from its young. While doing so, it continues to monitor the movements of the intruder, while keeping track of the location of its young. Based on this information, the plover may adjust direction to insure the intruder moves away from the young.[10]

---

8 However, Kornblith goes on to argue for the appropriateness of full-blown knowledge attribution. In brief, he argues that 'knowledge ... first enters our theoretical picture at the level of understanding of the species, rather than the individual' (2002, 57). While the behaviour of a specific individual may be explained on the basis of intentional states other than knowledge, at the level of species it must be explained how 'members of the species are endowed with a cognitive capacity that allows them successfully to negotiate their environment' (2002, 57). Kornblith argues that explanation of the role of adapted cognitive mechanisms in the generation of successful action on the part of an animal requires the attribution of knowledge.

9 Such an analysis of the plover would have much in common with the kind of psychologically embedded epistemology that Quine describes in his (1969, 82-3).

10 The possibilities for epistemological analysis do not stop here. For example, it might further be shown that plovers acquire knowledge in a variety of different ways (e.g., non-

It would be fair to describe the outcome of such an analysis as an epistemology of the piping plover. In order to produce such an epistemology we need to occupy a vantage point external to that of the plover itself. Yet it seems entirely possible to describe the plover's epistemic situation from a perspective outside of the plover's own point of view.

This has interesting implications for our own case. For there is no apparent reason why we should be unable to carry out an epistemological analysis of the kind just described for the plover with respect to ourselves. To do so, we might proceed in a manner not unlike Ristau's investigation of the piping plover. We can determine how humans acquire knowledge about their environment by means of their senses and reasoning processes. We can explore the reliability and limits of our senses and reasoning by means of the scientific study of perception and inference. By investigating the relation between human knowledge and behaviour, we can explain how our knowledge enables us to successfully negotiate our environment.

In thus developing an epistemological model of ourselves, it is true that we must turn our gaze upon ourselves. We must regard ourselves from our own point of view. But it is not clear why the ability we have to carry out an epistemological analysis of the plover should suddenly desert us when we attempt such an epistemological analysis of ourselves. There seems no reason in principle why we should be unable to conduct an empirical investigation of our own epistemic capacities in a manner analogous to that employed in the case of the piping plover.

But, if this is so, then it seems clear that we are able to adopt a viewpoint external to ourselves. For we may conduct an epistemological study of humans on the basis of which we are able to explain how human epistemic states give rise to successful interactions with our environment. In so doing, we adopt the perspective of an external observer of our own human epistemic situation. From within such a perspective, we are able to propose an epistemological model of the relation between human thought and our surrounding environment. Such a model may, of course, be prone to error, and is anything but certain. However, this does not show that it is impossible to adopt such a perspective. It only shows that the result of adopting such a perspective need not be an infallible view of the world.

Perhaps, in the end, this is all that Putnam's claim comes to. We are unable to adopt a point of view from which to gain infallible insight into the way of the world. So we cannot adopt a God's Eye point of view. For, while we may take up an external vantage point with respect to our own epistemic situation, we do not know everything and we may be mistaken. But this should be no surprise to anyone. After all, we are not God.

Appendix: On 'The Limits of Realism'

In a recent paper entitled 'The Limits of Realism', Michele Marsonet has drawn on considerations similar to Putnam's concerns about the God's Eye point of view to

---

visual sensory modalities) or that such knowledge is subject to certain limitations (e.g., due to eye placement or lack of night vision).

argue that realism is subject to significant limitations. These limitations are primarily due to the role played by conceptual schemes in our cognitive activities:

> As [far] as humans are concerned ... the world is characterized by a sort of 'ontological opacity' which makes the construction of any absolute ontology very difficult. Our ontology is characterized by the fact that the things of nature are seen by us in terms of a conceptual apparatus that is inevitably influenced by mind-involving elements. (Marsonet, 2002, 190)

While Marsonet grants the existence of an independent reality, he asserts that we 'have access to ... [mind-independent] things only via [our] conceptual apparatus' (2002, 193). The point is not that concepts constitute 'natural reality', but that we 'perceive this same reality by having recourse to the filter of a conceptual apparatus' (2002, 191). The result is that there is no 'clear distinction between ontology and epistemology' (2002, 191). Truth is 'essentially tied to human interests' (2002, 190) and is 'essentially "relative"' (2002, 194).

Marsonet's paper is a synthesis of a range of arguments against realism that arise from reflection on the role of concepts in our thinking about the world. Marsonet is concerned with realism in general, rather than with scientific realism in particular. But his arguments are of clear relevance to the position of scientific realism as I understand it. Rather than address all of the issues raised by Marsonet's paper, I will respond to the points that appear to be especially central to his overall line of argument.

Let me first address the question of whether it is possible to draw a line between ontology and epistemology. This is not the question of whether we can distinguish between two subject areas in philosophy, on the one hand, the area that studies what exists, and, on the other hand, the area that studies the nature of knowledge. Rather, Marsonet's point is that we cannot specify what exists without using concepts that reflect the contribution of the human mind. As he says, 'if our conceptual apparatus is at work even when we try to pave our way towards an unconceptualized reality, our access to it inevitably entails the involvement of the mind' (2002, 191). In other words, any attempt to specify the nature of mind-independent existence must inevitably fail to do so, since it requires the use of human concepts which are mental in provenance.

I will turn to the problems posed by concepts shortly, but would like first to make two general remarks about the relation between epistemology and ontology. First, by granting that there is an independently existing reality Marsonet has granted all that the realist requires from an ontological point of view. For he grants that reality does not depend on the knowing subject. All that remains at issue is the question of whether the independently existing reality is a knowable reality. Second, while the realist denies that the knowable world depends on the knowing subject, it is not clear that the realist need insist that questions of epistemology are fully independent of questions of ontology. As I will argue later, the realist must draw upon facts about the way the world is to explain how we manage to arrive at knowledge of the way the world is (see Chapter 8).

Turning to the problem of conceptual schemes, let me begin with Marsonet's claim about the mind-involving character of our concepts. As we have just seen, Marsonet takes this to entail that we are only able to represent 'unconceptualized reality' by means of concepts which reflect the influence of our minds. Thus, we are strictly unable to represent unconceptualized reality as such. But to conclude from the mind-involving character of concepts that we are unable to form a conceptual representation of a mind-independent reality appears to trade on a fallacious view of the relation between our minds and the content of our concepts. To have a concept one must have a mind. But the fact that having a concept depends on having a mind does not entail that the concepts that we have are concepts of mind-dependent things. Our concepts may be concepts of mind-independent things despite being our concepts. Thus, we may form the concept of a thing that exists independently of the mental, the content of which makes no reference to the mind, even though forming the concept is itself a mental activity and the concept is an object of thought. I can therefore see no basis for the conclusion which Marsonet appears to draw that we are unable to form the concept of a mind-independent reality (or of mind-independent things) because our concepts reflect 'the involvement of the mind' (2002, 191).

As for the question of the extent to which our knowledge of the world is undermined by our need to use concepts, let me start by quoting at greater length from a previously cited passage:

> No one denies that it would be good to transcend our conceptual machinery in order to glimpse at how the world really is, independently of any view we can hold about it. This, however, cannot be done because of the very way we are made. Unlike some forms of classical idealism, we can recognize the presence of things that are 'real' in the sense of being mind-independent but, on the other hand, a qualification is needed to the effect that human beings have access to those things only via their conceptual apparatus. (2002, 193)

In this passage, I take Marsonet to suggest that it is impossible for us to have direct epistemic access to the world independent of conceptual scheme. We are unable to adopt a standpoint devoid of concepts from which to gain conceptually unmediated access to the world. Instead, we must view the world through the filter of some conceptual scheme. Thus, Marsonet appears to conclude, or at least strongly suggest, we cannot know the world as it really is, in its own right, separate from our conceptual contribution.

But no such sceptical conclusion about our inability to have knowledge of reality follows from the fact that we must think about the world in terms of some set of concepts or another. The mere fact that we must think in terms of concepts does not entail a sceptical conclusion about our inability to have epistemic access to the world independent of concepts. This would only follow if all of our concepts were necessarily mistaken representations of the world. But the mere fact that we must use concepts does not entail that our concepts are mistaken. They might, in fact, be correct representations of the world. And so we could, in principle, know the way the world really is, despite the fact that we must represent the world to ourselves using our conceptual representations of the world.

Of course, it might be replied that, given our inability to step outside all of our concepts, we can never know whether our concepts do provide an accurate depiction of reality. But this reply is surely specious. There is no reason, short of adopting an extreme form of scepticism, to assert that all of our concepts fail to represent reality. To suppose, for example, that our ordinary everyday concepts fail to correspond to anything real leaves us without a viable explanation of the success of our practical interactions with the world. And it renders mysterious our apparent ability to engage in communicative exchange with other human beings. In short, it is not clear that extreme scepticism about conceptual representation is even an intelligible hypothesis. It is therefore safe to assume that, at least in large part, our basic conceptual apparatus is a reasonably accurate representation of reality.

But there is no need to rest content with a brusque dismissal of extreme scepticism. For it is possible to provide a positive rationale for a robust realist view of conceptual representation. It can, for example, be argued, as it has been by Hilary Kornblith, that the best explanation of the success of science is that our natural kind concepts are by and large accurate representations of the real natural kind structure of the world (see Kornblith, 1993a, 40-47, 74-8, 105-6). This accounts, not only for the success of science, but for the reliability of our inductive reasoning in both ordinary and scientific contexts. Kornblith's application of the success argument to our conceptual apparatus can be used to turn the issue back on Marsonet. For if Marsonet wishes to deny that our concepts are more or less accurate portrayals of a mind-independent reality, the question arises of how humans are able to engage in systematically successful action in both everyday and scientific affairs, if indeed all of our concepts fail to represent the world. Surely, such success would be an inexplicable miracle if our conceptual schemes are systematically mistaken.

The problem of conceptual schemes does not, therefore, pose a genuine threat to realism. It remains to pass brief comment on Marsonet's apparent acceptance of a relativist view of truth. Marsonet is struck by the fact that both our commonsense and scientific beliefs about the world are subject to ongoing variation. Thus, he suggests, truth is 'essentially "relative" and bound to evolve with the passing of time' (2002, 194); 'relativism and fallibilism are not ghosts to be afraid of, but just inevitable factors of our relationship with the surrounding environment' (2002, 200). While I agree wholeheartedly with the point about fallibilism, it is a mistake to infer relativism about truth from the fact that our beliefs about the world are subject to variation. This is to conflate what is *accepted* as true with what *is* true. While the former may well vary, the latter need not. What vary from time to time are the beliefs that humans happen to hold with regard to the world. But nothing about such variation of belief entails that the truth of such beliefs is relative to context. Given the independence of the reality that makes our beliefs true, a great many of our past beliefs about the world should simply be regarded as having been false, rather than to have been true in some sense relative to historical context. Thus, fallibilism, not relativism, is the lesson to be drawn from variation of belief.

Chapter 3

Truth and Entity Realism

3.1 Introduction

Much is made in some circles of a distinction between truth-orientated and ontological versions of scientific realism. In this book, I seek to defend a version of realism committed to both the pursuit of truth and the reality of the entities of science. However, I regard ontological versions of realism as legitimate versions of the doctrine. In this chapter, I seek to show that a genuine distinction exists between ontological versions of the doctrine and more usual forms of scientific realism such as that defended here.

In particular, I will examine the role played by the notion of truth within the version of scientific realism that is known as entity realism. Entity realism is the thesis that the unobservable theoretical entities of science are real. It is an ontological thesis about the existence of theoretical entities.

By contrast, scientific realism is often characterized as a thesis primarily involving the truth of theories. Sometimes scientific realism is expressed as the thesis that theoretical statements are intended as true descriptions of reality. Another favoured theme is that theoretical statements are objectively true or false in virtue of the way the world is independently of us. To such formulations it is usually added that the sense of 'true' required by scientific realism is the correspondence sense. To mark the contrast with entity realism, I shall say that a formulation of scientific realism which employs the notion of truth is a semantic version of scientific realism.[1]

The question I will address is whether entity realism is a non-semantic thesis distinct from semantic versions of scientific realism. More specifically, does entity realism express or, without further assumption, entail a version of the thesis of scientific realism which involves the notion of truth?

3.2 Entity Realism

When a doctrine is labelled 'realist', this is in general because the doctrine lays claim to the reality of some entity or kind of entity. In the case of entity realism, what is claimed to be real are the unobservable entities postulated by scientists (e.g.,

---

1 Here I follow Michael Devitt in describing a version of scientific realism cast in terms of truth as semantic. Devitt (1991, 39) takes a construal of realism which makes use of the notions of truth or reference to be a semantic thesis. Thus, while the notion of truth is of most present relevance, a version of scientific realism is semantic if any semantic notion occurs in its formulation.

electrons, atoms) which are usually referred to as 'theoretical entities'.[2] As such, entity realism is opposed to doctrines such as phenomenalism and instrumentalism, which reduce theoretical entities to experience or regard them as fictions.

To say that theoretical entities are real is not just to say that they exist. For a phenomenalist who holds that theoretical entities are to be construed as logical constructs out of experience may claim that, so construed, such entities exist. To exclude such construals, something further must be said about the existence of theoretical entities. According to entity realism, theoretical entities exist in their own right, without in any way depending on human thought or experience. They enjoy a mind-independent mode of existence which, contrary to phenomenalism, is not reducible to being the object or content of experience.

Entity realism is therefore the doctrine that theoretical entities are real in the sense that they exist mind-independently. This may be expressed more simply as follows:

(ER) The theoretical entities postulated by science are real.

ER captures the idea that theoretical entities are neither phenomenal constructs nor mere predictive devices.

It is important to note three qualifications about this formulation of entity realism. First, not all of the theoretical entities to which scientists purport to refer are real. Some theoretical entities which scientists assert to exist turn out not to exist. Second, not all apparent discourse about theoretical entities should be taken to imply genuine ontological commitment. Scientists sometimes employ idealizations (e.g., perfect gases, frictionless planes) which are not meant to be taken literally as accurate descriptions of real things. Third, entity realists do not assert the existence of all entities postulated by scientists, but only those entities for which there is good evidence. In order for ER to accurately reflect entity realism, it would need to be reformulated in light of these qualifications.

But, for present purposes, there is no need to modify ER. This is because the question to be pursued here is whether realistic treatment of theoretical entities engenders a thesis concerned with truth. Such a question is strictly concerned with the semantic implications of treating such entities as real. It may well be the case that realistic treatment is not warranted in the case of all theoretical entities. Yet the question of whether there are limitations on the extent to which realistic treatment of entities is justified is a separate issue from the question of the semantic implications of treating such entities as real. Thus, the latter question may be considered independently of the limitations entity realists may wish to impose on ontological commitments to theoretical entities.

---

2 A distinction may be made between those entity realists who are realists about the unobservable entities postulated in theoretical science and those who are realists about the unobservable entities postulated in the context of experimental science. For present purposes, I shall overlook this distinction, treating both as entity realists. The crucial issue is whether entity realism is a non-semantic thesis, not whether theoretical entities are postulated by theorists or experimentalists.

It might further be objected that since ER concerns the *postulation* of entities by science, it is therefore a semantic thesis. For if the entities postulated by science are real, as ER suggests, then the existence claims made by theories in postulating such entities are true. Thus, a semantic thesis about the truth of the existence claims of theories does follow from ER. However, as I will argue in Section 3.5, what follows is a weak semantic thesis, which is not committed to any particular conception of truth.

3.3 Is Entity Realism a Thesis about Truth?

In this section I will discuss the relation between entity realism and truth. My point of departure will be Michael Devitt's suggestion in *Realism and Truth* that realism is not a semantic doctrine. Devitt defines realism as follows:

> Tokens of most current common-sense and scientific physical types objectively exist independently of the mental. (1991, 39)

He then considers whether realism expresses a semantic thesis about truth:

> What has truth to do with Realism? On the face of it, nothing at all. Indeed, *Realism says nothing semantic at all* beyond, in its use of 'objective', making the negative point that our semantic capacities do *not* constitute the world. (1991, 39)

On the face of it, Devitt is right to claim that realism says nothing semantic. For realism, as defined by Devitt, is an assertion about the existence of certain entities. It is therefore an ontological thesis about the existence of things rather than a semantic thesis.

I will now elaborate Devitt's suggestion in the context of the earlier formulation of entity realism ER. Consider, first, whether ER says anything semantic. On the face of it, as Devitt might say, it does not. ER says only that some non-linguistic things are real. It is neither formulated in a metalanguage as a thesis about expressions of an object-language, nor does it apply a semantic predicate such as 'true' to any item capable of bearing semantic properties. In the absence of semantic and metalinguistic devices, ER would appear not to express a semantic thesis.

This alone may seem enough to establish that ER is a non-semantic thesis distinct from any semantic version of scientific realism. However, it might be objected that it is a trivial task to produce an alternative formulation of ER that is a semantic thesis. For the device of semantic ascent guarantees that any object-linguistic assertion has an equivalent metalinguistic formulation cast in terms of truth. In particular, semantic ascent on ER yields the following statement:

> (ER*) The statement 'The theoretical entities postulated by science are real' is true.

ER* is itself true if and only if the entities postulated by science are real; that is, if and only if ER is true. It is couched in a metalanguage and it contains the semantic term 'true'. Thus ER is equivalent to ER*, and ER* is a semantic thesis. It would seem to follow that ER is semantic.

This objection depends on the precise sense in which an object-linguistic assertion and the associated metalinguistic assertion of its truth are equivalent. Since such assertions agree in truth-value, they are at least materially equivalent. But mere material equivalence of ER and ER* hardly shows ER to be a semantic assertion. For statements the same in truth-value may differ radically with respect to content. Thus ER and ER* might be materially equivalent even though one is a semantic thesis and the other is not. Therefore, in order to establish that ER is semantic, the objection requires a stronger form of equivalence.

In fact, the truth-value of an assertion is necessarily the same as that of an assertion of its truth. It is not possible for one to be true and the other false. Thus, the relation between such assertions is indeed stronger than material equivalence. Since it is impossible for such assertions to diverge in truth-value, their relation is one of logical equivalence. However, even the fact that ER and ER* are related by this stronger form of equivalence does not establish that ER expresses a thesis of the same kind as ER*. For, as I will now argue, despite the logical equivalence of ER and the semantic thesis ER*, ER is not a semantic thesis.

In the first place, it is a mistake to infer from the logical equivalence of ER and ER*, and the occurrence of 'true' in ER*, that ER is a semantic thesis. For this would imply that any statement whatsoever is semantic, since for any statement whatsoever there is a logically equivalent formulation of it which is an assertion of its truth. But to say that any statement whatsoever is semantic trivializes the idea of being semantic, and completely removes the point of describing something as a semantic thesis.

Against this, it might be countered that a pragmatic implication of the act of asserting a statement is that the statement asserted is put forward as true. All assertions are therefore implicitly semantic, since they assert the truth of a statement. But the question here is not whether the act of asserting a statement implies that it is put forward as true. Rather, the question is whether the content of the statement is itself a semantic thesis which involves in some way a semantic notion such as truth.

The main problem with the above argument that ER is semantic is that it rests on the assumption that logically equivalent assertions have the same meaning. Specifically, it assumes that because ER and ER* are logically equivalent they express the same thesis. The trouble is that logically equivalent assertions are not necessarily semantically equivalent. Assertions may be such that there is no possible condition in which one is true and the other false, yet they may differ in meaning.

In particular, it does not follow from the fact that ER and ER* necessarily converge in truth-value that they mean the same thing. ER is a statement about the theoretical entities of science, whereas ER* says that that statement is true. But to say that a statement is true is to assert something about the statement which is not expressed by the statement itself.[3] It is to say *of the statement* that it has a particular semantic property, namely truth. But to say of a statement that it is true is to say

3 This is to reject the claim made by some redundancy theories of truth that '"P" is true' means the same thing as 'P', a claim which is difficult to sustain given that the former is about a sentence while the latter is not. For the point that the redundancy theory denies that '"P" is true' is a statement about 'P', see Horwich (1990, 39).

something different in meaning from saying of some entities that they exist. Given this lack of synonymy between ER and ER*, the entity realist's ontological thesis is distinct from the semantic thesis derived from it by semantic ascent. ER is not a semantic thesis.

There is a further problem with use of semantic ascent to show that ER is semantic. If the logical equivalence of ER and ER* does not entail that ER is semantic, the only way semantic ascent can show ER is semantic is if it transforms its meaning into something semantic. But sentences are meaning invariant with respect to assertion of truth. That is, assertion of the truth of a sentence does not result in transformation of the meaning of the sentence of which truth is asserted. Thus, while semantic ascent on the entity realist's thesis yields an assertion of its truth, it does not alter its meaning. In particular, it adds nothing semantic to the statement of the thesis itself. So, despite semantic ascent, the meaning of ER once truth is attributed to it by ER* remains precisely the same, namely, a statement of the reality of certain entities.

3.4 Truth and Existence Claims

In the previous section I argued that ER is neither rendered semantic by virtue of logical equivalence to ER*, nor transformed into a semantic thesis by semantic ascent. I conclude from this that semantic ascent on ER does not show it to express a semantic thesis such as ER*. However, while ER may not express a semantic thesis, it might seem to entail one. I now turn to the question whether ER entails a semantic thesis without further assumption.[4]

Consider the thesis expressed by ER that the theoretical entities postulated by science are real. What follows if a particular entity postulated by a given scientific theory is indeed real? If a theory postulates the existence of an entity which in fact exists, what the theory says when it postulates that entity is true. While the theory may otherwise say some false things about the entity, the theory's existence claim regarding the entity is true, provided only that the entity exists.

In this way, it appears that the entity realist's ontological thesis concerning the existence of theoretical entities leads to a semantic thesis concerning the truth of the existence claims of theories. For it follows from the thesis that the theoretical entities postulated by science are real that the existence claims which postulate theoretical entities are true. In other words, ER entails

(ER**) Scientific claims which postulate the existence of theoretical entities are true.

ER** refers to linguistic items and predicates truth of those items. So it states a semantic thesis.Since ER** is an apparent consequence of ER, ER seems to entail a semantic thesis after all. But let us look more closely at the inference from ER to ER**. The inference proceeds from the existence of theoretical entities, which ER asserts, to the truth of the existence claims made by theories in postulating such

4 Of course, given the logical equivalence of ER and ER*, the truth of ER entails the truth of a semantic thesis, namely, ER*. But to say that the truth of ER entails that of ER* is not to say that one can be derived from the other without further non-logical assumptions.

entities, which ER** asserts. Such an inference requires an assumption about the relation between existence and the truth of existence claims. In particular, it assumes that if an entity exists, then the sentence which asserts its existence is true. But this is not a special assumption about the relation between existence and the truth of existence claims. Rather, it is a simple instance of semantic ascent from an object-linguistic sentence to an assertion of the truth of the sentence in a metalanguage.

Thus, while the thesis, ER**, that the existence claims of theories are true is indeed a semantic thesis, it is not a thesis that follows from entity realism without further assumption. Rather, ER** is a consequence of ER which follows from ER only by means of semantic ascent. Strictly speaking, therefore, the entity realist's thesis of the reality of theoretical entities does not immediately entail a semantic thesis about the truth of theories.

3.5 Semantic Ascent and the Theory of Truth

Given semantic ascent, ER entails a semantic thesis which involves the notion of truth. It may now seem a short step from entity realism to a standard version of scientific realism which incorporates a substantive conception of truth, such as a correspondence theory. However, semantic ascent is not tied to any particular theory of truth. Thus, as I will now argue, any semantic implication arising from entity realism via semantic ascent is similarly uncommitted with regard to the nature of truth.[5]

Let us first consider the basis of semantic ascent. The device of semantic ascent is derived from the standard disquotational truth schema, or T-scheme:

> 'P' is true if and only if P.

To ascend semantically from the assertion of a sentence to the assertion of the truth of the sentence is to proceed in accordance with the conditional 'If P, then "P" is true'. Since this conditional follows from the T-scheme, semantic ascent is licensed by the T-scheme.

The trouble is that views diverge widely on the relation between the T-scheme and the concept of truth. While there are those who regard the T-scheme as a definition of truth, others regard it as a minimal constraint which any complete concept of truth must satisfy. Thus, to endorse the T-scheme is not to endorse any particular theory of truth.

Semantic ascent suffers from a similar ambiguity. It takes over from the T-scheme the same imprecision about the nature of truth. As a result, the semantic thesis derived from entity realism via semantic ascent does not specifically commit

5 It should be noted that considerations similar to those of the preceding section reveal that ER leads to a semantic thesis concerning reference. Roughly, if a given theoretical entity is real, then a term purportedly referring to it has reference. However, given the variety of theories of reference available, it can be shown by an argument analogous to the one to be given in the present section that the resulting thesis is a minimal semantic thesis committed to no particular conception of reference.

entity realism to any particular theory of truth. So, while it is indeed a thesis about the truth of certain theoretical claims, there is no indication what truth is.

To see this, it is sufficient to note that the T-scheme is common ground to all the standard theories of truth. The disquotational, pragmatic, coherence, internalist, verificationist and correspondence theories of truth all agree that truth, *whatever it is*, must conform to the T-scheme. The disquotationalist claims that there is no more to truth than disquotation, so that the T-scheme exhausts the meaning of the word 'true'. The remaining truth-theories differ on what is to be added to the T-scheme in order to fully specify what 'true' means. The pragmatist says to be true is to be useful, the coherence theorist that it is coherence with a system of beliefs, the internalist that it is what is accepted in the ideal limit of science, the verificationist that it is to be verified, and the correspondence theorist says that it is a relation between sentences and extralinguistic items.

Since semantic ascent derives from the T-scheme, and the T-scheme is common ground to the major truth-theories, it follows that semantic ascent is also common ground to the major truth-theories. Given this, the semantic thesis derived from entity realism constitutes an uninformative thesis about truth. Since it is arrived at by means of the device of semantic ascent, which is common to all the best-known theories of truth, it is committed to no particular theory of truth.

3.6 Three Objections

In light of the neutrality of the T-scheme, the semantic stance of entity realism seems minimal indeed. For, beyond complying with the T-scheme, the semantic thesis to which it gives rise via semantic ascent is perfectly neutral with regard to the nature of truth. However, the idea that entity realism combined with semantic ascent yields no specific interpretation of truth is bound to be controversial. In this section I will consider three objections that are likely to arise.

Objection one: Entity realism is not consistent with all theories of truth. For example, it is inconsistent with the coherence theory, which makes existence contingent on relations between beliefs, in violation of the mind-independence aspect of realism.[6] As a result, it is impossible to combine all truth-theories with entity realism, which is not therefore neutral with respect to theory of truth.

Reply: The point of my argument in the last section was not that the entity realist can embrace any theory of truth at all. The point, rather, was that to embrace both entity realism and semantic ascent is not yet to embrace any particular theory of truth. Absent further argument, a gap remains between semantic implications of entity realism derived via semantic ascent and versions of scientific realism which incorporate a substantive conception of truth.

Objection two: If entity realism is inconsistent with certain truth-theories, it cannot be denied that entity realism is a semantic thesis which involves a definite

6 The point here is the general point that realism cannot be combined with an epistemic theory of truth. For further discussion, see 7.6 and 8.2.

view of truth. For, unless entity realism says something specific about truth, it cannot disagree about the nature of truth with any theory of truth.

Reply: The inconsistency between entity realism and some theories of truth is not due to disagreement about the nature of truth. It is due, rather, to conflict between a consequence of certain truth-theories with respect to the existence of entities and the mind-independence aspect of entity realism. That is, the conflict is due, not to what entity realism says about truth, but to what a truth-theory says about existence. Thus it does not follow from the inconsistency of entity realism with a truth-theory that entity realism is committed to any specific conception of truth.

Objection three: Entity realism requires a correspondence theory of truth. This is because of the commitment of entity realism to the mind-independent existence of theoretical entities. Given the contrast between mind-independent entities and theoretical existence claims couched in language, the truth of such claims must consist in a correspondence relation between linguistic and extra-linguistic items.

Reply: It is certainly the case that, if entity realism is to be combined with a theory of truth, a theory of truth is required that is consistent with the mind-independent existence of theoretical entities. But while this requirement may reduce the range of truth-theories consistent with entity realism, it need not reduce the range of suitable candidates to the correspondence theory alone. Theories of truth which tie truth closely to epistemic evaluation fall foul of the independence aspect of entity realism. However, there exist theories of truth apart from the correspondence theory, which do not identify truth with an epistemic property, and which are therefore consistent with mind-independence.[7]

Moreover, even if entity realism did require a correspondence theory, it would not follow that 'true' occurs in its semantic consequence in a correspondence sense. The semantic consequence is generated from entity realism by semantic ascent. Semantic ascent is subject only to the constraint of the T-scheme, which is neutral with regard to theory of truth. Therefore, the term 'true' occurs in it as a term that is neutral between the various truth-theories.

3.7 Conclusion

In this chapter I have considered the question whether entity realism is a thesis about the truth of theories. I rejected the attempt to render it semantic by embedding it within an assertion of its own truth. I argued that semantic ascent on the existence claims of theories yields a thesis concerning the truth of such claims. I then noted that this semantic thesis does not follow from the statement of entity realism without further assumption. I observed that the resulting semantic thesis is noncommittal about the nature of truth. Finally, I defended this observation against three objections which suggest that entity realism has specific truth-theoretic commitments.

Entity realism is not to be identified with a semantic thesis. It does not itself assert, nor does it immediately entail, a thesis about the truth of theories. Moreover,

7 There is, in general, no conflict between the mind-independence aspect of entity realism and deflationary theories of truth, such as disquotationalism, the redundancy theory, and Horwich's minimalism.

while a thesis involving truth does follow from entity realism by means of semantic ascent, the entity realist is not thereby committed to any specific account of truth. Certainly, the minimal semantic thesis derived via semantic ascent does not itself provide support for a realist seeking to incorporate some specific truth-theory within scientific realism. Such a realist must find grounds independent of entity realism for embracing any particular theory of truth.

I conclude, therefore, that entity realism is a genuine non-semantic alternative to truth-orientated versions of scientific realism. But, while I grant that entity realism is a distinct form of realism from such semantic formulations of the doctrine, I remain optimistic that a robust defence of a truth-orientated version of scientific realism may be provided. That is the task pursued in other chapters of this book.

Chapter 4

Incommensurability and the Language of Science

4.1 Introduction

For much of the past century, the philosophical imagination has been captivated by language. Few aspects of language, or of language use, have escaped philosophical scrutiny. Traditional philosophical problems have been recast as issues in the philosophy of language. And philosophers have employed techniques of linguistic and conceptual analysis in an attempt to solve or dissolve these problems.

The philosophy of science has been no exception. From the topics of confirmation and explanation to those of laws of nature and the dynamics of theory change, linguistic aspects of science have loomed large. One enduring theme has been an interest in the semantic and epistemic features of scientific discourse about the world. Here two key questions have been the focus of much attention. First, how does the vocabulary used by scientists acquire its meaning? Second, how is scientific vocabulary related to reality? One might characterize these two questions, respectively, as the question of the *meaning* of scientific terms as opposed to the question of the *reference* of such terms.

Debate about these two questions has in large part been conditioned by a distinction between two kinds of vocabulary employed in science. On the one hand, there is the observational vocabulary (e.g., 'red', 'smooth'), which scientists employ to report upon observable phenomena and objects that are directly perceived by means of their senses. On the other hand, scientists employ theoretical vocabulary (e.g., 'electron', 'gene') when they speak about the unobservable entities which are postulated by scientific theories to explain observable phenomena. The distinction turns, at base, on a contrast between observable and unobservable entities. But it also reflects a difference in the conditions which govern the application of the two kinds of vocabulary. For, while observational terms may be applied on the basis of immediate experience of the items to which they refer, theoretical terms are unable to be employed on the basis of direct perception of their referents.

For philosophers of a broadly empiricist persuasion, the distinction between observational and theoretical terms marks an important semantic and epistemic difference. This is principally because observational terms may be defined, as well as applied, on the basis of experience alone. It is also due to the fact that the truth of empirical claims, which employ observational terms to describe observed phenomena, may be established by means of direct empirical test. By contrast, neither the meaning of a theoretical term, nor the truth of a claim about unobservable theoretical entities, may be established on the basis of observation alone.

For philosophers less inclined to empiricism, the distinction between observational and theoretical terms bears little significance, and may even seem misconceived. It has, for example, been argued by some philosophers that the distinction fails to correspond to a genuine difference in linguistic use, since theoretical terms may be applied to observable items, and vice versa. It has also been argued that, regardless of the validity of the distinction, scientists routinely describe observed phenomena using the terms of an operative scientific theory. Most importantly for present purposes, it has been argued that the meaning of observational terms depends upon theoretical context. Thus, while there may be a pragmatic distinction based on the conditions of application of terms, no semantic distinction may be drawn between observational and theoretical terms.

The idea that the meaning of observational vocabulary depends on theoretical context has serious implications for the rationality of scientific theory choice. For if the meaning of observational terms depends on theoretical context, rather than being fixed independently of theory, then the meaning of such terms may vary with theory. In the absence of a semantically invariant observation language, which is neutral between theories, the problem arises of how to compare the empirical claims which alternative scientific theories make about the world. If there is no shared observation language, then what one theory says about the world may neither agree nor disagree with any empirical claim made by an alternative theory. But if claims made by alternative theories cannot be compared, then there may be no rational basis on which to choose one theory over the other. The thesis that the content of theories may be incomparable for semantic reasons is known as *the incommensurability thesis*.

The issue of meaning variance interacts in complex ways with central currents in 20th century philosophy of science. In this chapter, I employ meaning variance as a unifying theme to trace some major developments in philosophical accounts of the language of science which have occurred in the past century. In Section 4.2, I briefly examine the empiricist account of scientific discourse associated in the first half of the 20th century with the philosophical movement of logical positivism. Sections 4.3 and 4.4 trace the problem of meaning variance to the emergence in the second half of the century of a post-positivist or historical school in the philosophy of science. Sections 4.5 and 4.6 consider the attempt to treat the problem as a problem about reference, with special emphasis on the use made by scientific realist philosophers of science of a causal theory of reference. In Section 4.7, I briefly sketch the approach based on a causal-descriptive account of reference that I have proposed as a response to meaning variance and incommensurability. Then, in Section 4.8, I examine the implications for realism of the thesis of taxonomic incommensurability which is found in the later writings of T.S. Kuhn. Finally, in Section 4.9, I conclude by discussing the contention of Hoyningen-Huene, Oberheim and Andersen that my approach to incommensurability inappropriately relies on realism.

4.2 Logical Positivism

'Logical positivism' is the name commonly used for a scientifically minded philosophical movement which emerged in Vienna in the 1920s and 1930s. The movement grew up around a group of like-minded philosophers and scientists known as the Vienna Circle, which included such figures as Moritz Schlick, Otto Neurath and Rudolf Carnap, who met regularly to discuss philosophical ideas at the University of Vienna. Also associated with the Circle, but not members, were Ludwig Wittgenstein and Karl Popper, as well as the English philosopher A.J. Ayer, who visited Vienna in the early 1930s. With the rise of Nazism in the mid-1930s, the Circle dispersed, its members moving to the United States and Great Britain, where their influence significantly shaped the subsequent development of philosophy in the English-speaking world.

Logical positivism has usually been understood as an extreme form of empiricism.[1] As such, its supporters adhered to the view that both the content of our empirical concepts and the evidential support for our beliefs about the world must derive from sensory experience. The historical roots of logical positivism trace back to 18th century British empiricism, though the name 'positivism' derives from the *philosophie positive* of the 19th century French thinker Auguste Comte. What set the positivism of the Vienna Circle apart from its predecessors was a tendency to pose philosophical problems in terms of language, and a systematic attempt to apply the techniques of modern symbolic logic to the treatment of these problems.

I will now briefly discuss two central themes of logical positivism, which are of special relevance to the language of science: the principle of verifiability, and the meaning of theoretical terms.

The Principle of Verifiability

One central question addressed by logical positivism was the question of how cognitively significant discourse about the world is possible. The response favoured at first by positivists was that a significant assertion is one which may be tested for truth or falsity by means of experience. On the basis of such a response, positivists sought to enforce a sharp distinction between significant and non-significant discourse about the world. Since metaphysical claims about a transcendent reality lying beyond experience cannot be empirically verified, this led the positivists to dismiss metaphysics as meaningless nonsense.

The idea that significant assertion requires the possibility of empirical test is known as verificationism. Logical positivists expressed this idea in the form of the principle of verifiability, as exemplified by the following passage from Moritz Schlick:

> ... there is only one way of giving meaning to a sentence, of making it a proposition: we must indicate the rules for how it shall be used, in other words: we must describe the facts which will make the proposition 'true', and we must be able to distinguish them from the facts which will make it 'false'. In still other words: The Meaning of a Proposition is the Method of its Verification. The question 'What does this sentence mean?' is identical

1 Recent scholarship suggests that the usual understanding of logical positivism may be misleading. Far from being simply an extreme outgrowth of earlier empiricism, logical positivists were equally responsive to neo-Kantian philosophy. See, for example, Coffa (1991), Friedman (1993) and Parrini (1998).

with (has the same answer as) the question: 'How is this proposition verified?' (Schlick, 1938, 34)

The principle that meaning consists in method of verification serves two functions. First, as in the above quote, it specifies what the meaning of a sentence consists in, namely its conditions of verification. Second, it specifies a criterion of significance, which a sentence must satisfy in order to be meaningful. For, if no verification conditions are specified, then, as Schlick goes on to say, 'our words do not form a real proposition at all, they are mere noises without meaning' (Schlick, 1938, 35).

While many philosophers continue to associate meaning with verification, the strict identification of meaning with empirical verification conditions is now widely seen as untenable. As Schlick himself realized, meaning cannot require present testability, on pain of eliminating all currently untestable sentences as meaningless; it requires instead verification in principle. Nor may meaning be identified with conclusive verification, since, as Popper argued, that would render statements of the universal laws of science meaningless (Popper, 1959, 36). By the 1950s, the difficulties in applying verificationism at the level of sentences led C.G. Hempel to conclude that cognitive significance 'can at best be attributed to sentences forming a theoretical system, and perhaps rather to such systems as wholes' (1965, 117). Another problem that bedevilled positivism was the status of the principle of verifiability itself: since it is evidently not an empirical claim which may be verified by empirical test, the principle itself appears to be devoid of cognitive significance. While the positivists sought to meet this problem, for example by proposing the principle as a convention rather than a statement, some philosophers have taken it to show that positivism was fundamentally incoherent (e.g., Putnam 1981, 113).

The Meaning of Theoretical Terms

Another approach to cognitive significance pursued by logical positivism focused on the meaning of individual terms, rather than sentences. But, while the verifiability criterion made the meaning of theoretical statements problematic, similar problems arose as well for theoretical terms. For if experience is the source of meaning, terms whose meaning cannot be directly given by appeal to experience may fail to have any meaning at all.

Positivists tended to assume that observational terms acquire meaning by direct reference to observable entities, e.g. by ostensive definition. Since a theoretical term may not be defined by means of a direct, ostensive identification of an unobservable theoretical entity, it was thought that theoretical terms might be defined by means of a connection with observational vocabulary. Some early positivists held that the meaning of a theoretical term might be fully specified by means of an explicit definition using only observational terms. On such a view, a theoretical term is exactly synonymous with a complex expression consisting entirely of observational terms, and its sole function is as convenient shorthand for the complex expression. But explicit definition fails even before the level of full-blown theoretical terms is reached. As shown by Carnap in 'Testability and Meaning' (1936), dispositional terms (e.g., 'fragile', 'soluble') cannot be fully defined using observational terms alone. While the observational circumstances in which a disposition is displayed may be specified, the disposition itself obtains even when the observable circumstances do not. Yet the disposition *itself* (e.g., to dissolve if placed in water) is not something that can be observed.

In the absence of explicit definability, later positivists adopted a partial interpretation approach to the meaning of theoretical terms. In 'Testability and Meaning', Carnap showed how disposition terms may be introduced by means of 'reduction sentences' which specify the observable circumstances in which the disposition is manifest (e.g., 'If a glass is struck, then if it breaks it is fragile'). While reduction sentences do not fully define a disposition term, a set of such sentences partially defines the term by specifying empirical conditions in which it applies. Such specification of application conditions only partially defines the term because it fails to specify meaning for circumstances in which the observable conditions do not obtain (e.g., when the glass is not struck).

By the mid-1950s, the partial interpretation approach had come to form the basis of a liberalized positivist account of the language of science, known as the 'double language model'.[2] According to this account, the language of science divides into distinct observational and theoretical vocabularies, where theoretical terms are partially defined via 'correspondence rules' which link them to observational terms. However, while some theoretical terms derive their meaning via correspondence rules which link them directly to observation terms, others, having no such link, gain their meaning indirectly via links with other theoretical terms.

With the double language model, a holistic view of meaning starts to emerge. For on this account, the meaning of theoretical terms may depend on a variety of complex relations between observational and theoretical terms. As a result, the meaning of theoretical terms may be subject to variation with change of theory as the relations between theoretical and observational terms undergo revision in the course of theory change. In this way, the liberalized positivism of the 1950s allowed, at least in principle, for the possibility of meaning variance of theoretical terms. This possibility was not always recognized by advocates of the historical approach who sought to overcome positivist theory of language.[3]

2 For the double language model, see, for example, Carnap (1956) and Nagel (1961). As more liberalized forms of positivism appeared at the mid-century, it became increasingly common to refer to the approach by the name 'logical empiricism', rather than 'logical positivism', the latter sometimes being reserved for the more strictly empiricist approaches of the early part of the century.

3 A number of authors have noted the anticipation of the meaning variance thesis in positivist accounts of scientific language. See for example Reisch (1991), who discusses Carnap's letters to Kuhn, which Carnap wrote in his capacity as Editor of the *Encyclopedia of Unified Science*, in which Kuhn's *The Structure of Scientific Revolutions* appeared. One of the best early studies of the convergence between positivist theory of meaning and the meaning variance thesis remains English (1978).

4.3 The Historical Turn

In the 1950s and 1960s a new, post-positivist approach to the philosophy of science emerged, which highlighted developmental and contextual aspects of science, and placed great emphasis on the manner in which both scientific theories and scientific practice evolve as a historical process. The main participants in this 'historical turn' initially included such key figures as Paul Feyerabend, N.R. Hanson, Thomas S. Kuhn, Michael Polanyi and Stephen Toulmin, though they were later joined by Imre Lakatos and Larry Laudan. Of those initially involved in the historical movement, most drew considerable inspiration from the later philosophy of Ludwig Wittgenstein, Gestalt psychology and anti-Whiggish historiography of science. The result was an approach to science more inclined to interpret science in terms of its past, than to reconstruct it in terms of logic.

While historical philosophers of science divide among themselves on points of detail, it is possible to specify a number of salient themes which broadly characterize the historical approach as a philosophical school. It is typical of historical philosophers of science to emphasize the role and importance of large-scale, enduring traditions of scientific research. There is a tendency to reject a unique scientific method in favour of variation of methodological standards of theory-appraisal. No sharp division is made between theory and fact. Instead, observation is said to be 'theory-laden'. And the idea of an independently meaningful observation language is rejected in favour of the dependence of the meaning of observation terms on theoretical context. In what follows, I will briefly comment on each of these themes, though the final point will receive extended attention in the next section.

The importance of enduring traditions in the history of science was forcefully shown in the seminal text of the historical movement, T.S. Kuhn's *The Structure of Scientific Revolutions* (1970a), and further elaborated in subsequent work by Lakatos (1970) and Laudan (1977). These authors stress that individual scientific theories (e.g., Copernicus's or Kepler's theories of the solar system) are usually developed within the context of a set of underlying theoretical assumptions (e.g., heliocentric astronomy) which tend to be preserved through variation at the level of specific theory. Such underlying assumptions constitute a general perspective or world-view, and are the central ingredients in the deep-level theoretical frameworks which Kuhn described as 'paradigms'. They include both substantive assumptions about the way the world is and methodological ones about how to investigate the world. But unlike empirically testable predictions made by theories, such deep-level frameworks do not admit of empirical test, and may only be evaluated by comparison with rival sets of assumptions over a sustained period of time.

The issue of how to evaluate a scientific tradition points to the second major theme of the historical school. Traditional philosophers typically approached scientific theory appraisal in terms of a unique scientific method, employed throughout the sciences in all stages of the history of science. By contrast, the historical school laid great stress on the historical variation of patterns of scientific reasoning and theory appraisal. In their view, the history of science does not show that scientific method remains fixed while theories change, but that method is open to revision along with theory. Kuhn argued, for example, that standards of theory appraisal vary with scientific paradigm, and seemed to deny the existence of any such standards independent of paradigm (1970a, 94, 103). Somewhat later, Feyerabend argued that all methodological rules proposed by philosophers of science have in fact been violated at some stage in the history of science, and that there have been good grounds for violating them (1975, 23). Such denial of a fixed method has given rise to a widespread epistemological relativist denial of objective criteria of rationality amongst those whose work has been influenced by the historical movement.

One might, of course, think that even in the absence of a fixed method, sense experience might provide a neutral arbiter between rival theories. But it is just here, however, that the historical school comes into most direct conflict with traditional empiricist philosophy of science. For where empiricists held that sense perception provides objective, observer-invariant grounds for theory appraisal, historical philosophers of science tend to deny a sharp divide between matters of observed fact and theory. As was famously argued by N.R. Hanson, in his *Patterns of Discovery*, scientific observation is 'theory-laden', due to the thorough-going influence of theoretical background upon the content of experience. Thus, despite being presented with identical external circumstances, scientists may have divergent visual experiences, because, as Hanson says, 'there is more to seeing than meets the eyeball' (1958, 7). In similar vein, the phenomenon of theory-ladenness led Kuhn to suggest that 'the proponents of competing paradigms practice their trades in different worlds' (1970a, 150).

The theory-dependence of observation brings us back to language. For, it was argued not just that observation is theory-laden, but that the very language used to report observation depends on theory as well. As a result, rather than being independent of theory, the meaning of an observation term varies with the theoretical context in which it is employed. The issue of meaning variance leads to the topic of the next section, the semantic incommensurability of theories.

4.4 Incommensurability

One of the most controversial claims to emerge from the historical turn was the claim made by Kuhn and Feyerabend that alternative scientific theories may be incommensurable. While the meaning of the term 'incommensurable' resists precise specification, for present purposes it may be understood to mean that there are limits on the comparison of theories for evaluative purposes. More specifically, I will treat the claim of incommensurability as the claim that the content of one scientific theory may not be directly compared with the content of an alternative such theory.[4]

Kuhn set the claim of incommensurability within the context of his account of scientific theory change as revolutionary transition between paradigms. On Kuhn's account, the 'normal science' pursued by scientists under the guidance of a reigning

4 For a more nuanced discussion of the concept of incommensurability, see my (1997, Ch. 2), as well as Hoyningen-Huene (1993, 206-22), who offers an opposing analysis of the concept.

paradigm differs in fundamental ways from that undertaken within the context of an alternative paradigm:

> ... the normal-scientific tradition that emerges from a scientific revolution is not only incompatible but often actually incommensurable with that which has gone before. (1970a, 103)

The incommensurability of paradigms sets limits on the extent to which competing paradigms may be evaluated in a comparative manner, for example, by appeal to shared standards of theory appraisal, or conflicting predictions about the world. As a consequence, it is obscure on Kuhn's view how scientists may decide between competing paradigms on a rational basis.

For Feyerabend, the notion of incommensurability formed part of his critique of the empiricist model of the relations between successive theories (see Feyerabend, 1981, Ch. 4). Empiricists held that an earlier theory is either explained by, or reduced to, a later, more comprehensive theory by deductive subsumption of the laws of the earlier theory under the laws of the later theory. Feyerabend argued that the deductive relations between theories required by this model entail a semantic condition of meaning invariance, which is routinely violated in the course of scientific practice, as the meaning of scientific terminology undergoes profound changes in the transition between alternative theories.

It will simplify exposition to formulate the incommensurability thesis in terms of the following three features of the semantic relations between the vocabulary of alternative scientific theories: meaning variance, translation failure and content incomparability. The basic idea of the incommensurability thesis is that the content of alternative scientific theories is unable to be compared because of translation failure due to meaning variance of their vocabulary. I will now briefly discuss each of these points in turn.

The claim that meaning varies in the transition between theories may be presented by means of an argument which proceeds in three steps. The first step is to reject the empiricist assumption that there is a theory-neutral observation language which is meaningful independently of theory and semantically invariant between theories.[5] An independently meaningful observation language of the kind sought by empiricists would provide a semantically neutral medium of expression which would provide common ground between scientific theories, on the basis of which such theories may be directly compared. The second step is to argue, against the idea of an invariant observation language, that the meaning of observational terms depends on the theoretical context in which they are employed. According to such a 'contextual theory of meaning', an observational term obtains its meaning from the theory which is used to describe and explain the observed item to which the term is applied. Meaning is determined in a holistic manner, by means of the whole theoretical context in which the observational term is used. The third step

5 There are various ways to argue for such a rejection, e.g., by arguing against the ostensive definition model of language learning, or to argue that neither the experience nor the conditions associated with use of an observational term fix its meaning. The latter option is pursued by Feyerabend in his (1981, Ch. 2).

is to conclude that meaning varies due to the context-dependence of meaning. In particular, it follows from the context-dependence of meaning, that the meaning of observational and theoretical terms must vary with respect to the theoretical context in which they are employed. Thus, in the transition between scientific theories, there is change of meaning from one theory to the next. More precisely, where the same terms occur in different theories, such terms will be employed with a different meaning in the context of each theory.

It is important to mention two qualifications about meaning variance. First, the claim of meaning variance admits of two interpretations. *Radical* meaning variance occurs if all (or perhaps most) of the terms employed by theories change their meaning. *Partial* meaning variance occurs if only a restricted class of the terms change. The radical version tends to be found in Feyerabend, the partial version in Kuhn. Second, the sort of meaning variance of present relevance is restricted to alternative theories about the *same domain* of phenomena. Theories in different domains (e.g., continental drift and psychoanalysis) do not compete. Semantic variation between theories in different domains does not give rise to significant problems of theory comparison or choice.[6]

The claim that the vocabulary employed by theories semantically varies between theories does not, by itself, entail incomparability of the content of theories. The latter requires, beyond mere difference of meaning, that there be failure to translate between the vocabulary employed by meaning variant theories. There are a number of ways to argue for translation failure, of which perhaps the most straightforward is as follows. On the assumption that there is radical meaning change between theories, no term of one theory has the same meaning as any term of the other theory. As a result of such complete absence of common meaning, no statement entailed by one theory may be translated by means of a statement entailed by the other theory. Thus, there is a total translation failure between the theories.[7]

If meaning variance is understood to entail translation failure between theories, then a number of important consequences follow from meaning variance. For one thing, translation failure gives rise to difficulties of communication and understanding between the advocates of such rival theories. However, the consequence of most concern to philosophers of science is that the content of meaning variant theories is unable to be compared. That is, if it is impossible to translate from the vocabulary of one theory into another, then no statement of one theory can be matched with a

6 The requirement that meaning variant theories be theories about the same domain raises a serious difficulty for the incommensurability thesis. This is the so-called 'rivalry objection': how can theories with no meaning in common be rival explanations of the same phenomena? The objection applies with most force against claims of incommensurability based on radical meaning variance.

7 Of course, if the meaning variance is only partial, then the translation failure is also only partial. It should be added that a number of other ways of arguing for untranslatability have been employed. For instance, Feyerabend argues that the conditions of concept formation in one theory forbid the formation of concepts from another theory (1978, 68, fn. 118). Kuhn argues that within a theory a central complex of key terms is holistically inter-defined in such a way that terms from another theoretical complex are unable to be translated into it (2000, Ch. 2).

corresponding statement of the other theory which asserts or denies the same thing. There may be neither agreement nor disagreement between theories. But, if this is so, then it is impossible to directly compare the claims about the world made by one theory with the claims made about the world by the other theory. In short, the content of such theories is not directly comparable.

One immediate consequence of the inability to compare content is the impossibility of 'crucial experiments'. A crucial experiment is a test designed to choose between rival, equally well-supported theories, by means of a test of an observational prediction on which the theories disagree. If the result of the observation is unambiguous, such a test supports one theory while refuting the other. However, if no common language exists in which the consequences of rival theories may be expressed, then no predictive consequence of one theory may be formulated with which the other theory disagrees. Therefore, there may be no crucial test between such theories.

It is now possible to formulate the thesis of incommensurability in a perspicuous manner. Two alternative scientific theories are incommensurable just in case: (i) the meaning of the vocabulary employed by theories varies between theories; (ii) translation is impossible from the vocabulary of one theory into the vocabulary of the other; (iii) as a result of (i) and (ii), the content of such theories may not be compared.

4.5 Reference and Theory Comparison

The most influential objection to the incommensurability thesis has been the claim that, while alternative scientific theories may be meaning variant, they may still be compared with respect to content. The reason is that meaning variance between theories does not entail reference variance, and reference variance is what is needed for theories to be incomparable for content.

The referential response to incommensurability was forcefully elaborated by Israel Scheffler in his book *Science and Subjectivity* (1967, Ch. 3). Scheffler presented the response within the context of a Fregean distinction between sense and reference. According to this distinction, the meaning of a term divides into two components: the sense of a term is the concept or definition a speaker grasps when understanding what the term means, while the reference of a term is the object or set of objects which the term names.

Two features of the Fregean account of meaning are of most relevance in the present context. First, terms may differ in sense even though they refer to the same thing. Frege's classic example is that of the expressions 'evening star' and 'morning star', which have different senses though they refer to the same thing, viz., the planet Venus. Second, the sense of a term determines the term's reference. In particular, if we assume that the sense of a term is specified by means of a description, then the reference of the term is determined as that thing which satisfies the description which gives the sense of the term.

The crux of Scheffler's objection to the incommensurability thesis is that the content of theories may be compared by means of common reference, even if their terms differ in sense. The reason is that two sentences may be jointly incompatible though their constituent expressions do not have the same sense. Thus, suppose someone says 'The evening star is a star', while someone else asserts that 'The morning star is a planet'. Because the evening star and morning star are the same thing, and stars are not planets, it is impossible for both assertions to be true. Thus, despite employing words which differ in sense, the assertions are incapable of both being true, since they are about the same thing.

In general, statements about the world may conflict even if they contain different terms, or terms which differ in sense, provided that their constituent terms refer to the same things. Because of this, it is possible to compare the content of rival theories whose terms differ in sense. For, while rival theories might employ terms with different meanings, if their terms refer to the same things, then it is possible for their claims about the world to agree or disagree.

Reference Change

Despite the evident force of the referential objection, a reply to Scheffler's objection was readily available to Kuhn and Feyerabend. For they were able simply to argue that in cases of meaning variance between theories, both the sense and the reference of terms employed by theories are subject to change (cf. Kuhn, 1970a, 102, 1970b, 269; Feyerabend, 1981, 98). If they are right, Scheffler's appeal to the distinction between sense and reference is unavailing. For if reference is discontinuous between theories, there is no overlap of reference, and content cannot be compared.

There are a number of reasons to think there may be widespread change of reference between meaning variant theories. For one thing, historical cases suggest that change of reference is prevalent in the history of science; e.g., modern use of the term 'atom' seems not to refer to the same kind of entity as did ancient use of the term. For another thing, the differences in descriptions which theories give of the entities to which they refer may be so extreme that the descriptive content associated with terms must pick out completely distinct sets of things. Alternatively, the descriptive content associated with terms by one theory may be incompatible with the content associated with terms by another theory, so that the descriptive content of the alternative theories may not be satisfied by the same sets of things.

The crux of the issue is the assumption that sense determines reference. This is because the descriptions which scientists give of the entities they study are themselves subject to revision as scientists alter their theories about the world. But if descriptions of entities change in the transition between theories, the senses of the terms used by the theories will also change. Hence, there may be discontinuity of reference between theories. But without shared reference, the content of theories cannot be compared.

Thus, Kuhn's and Feyerabend's reply to Scheffler is that meaning variance includes reference as well as sense. Hence, an appeal to reference is unable to serve as basis for an objection to the incommensurability thesis. Moreover, if reference varies radically between theories, the content of theories may not be compared due to absence of common reference of theories. In this way, reference change gives rise to a referential version of the incommensurability thesis – incommensurability due to reference variance.

4.6 The Causal Theory of Reference

The claim by Kuhn and Feyerabend that reference varies in the course of scientific theory change is of particular concern to philosophers of a scientific realist persuasion. Scientific realists defend the view that the aim of science is to discover the truth about an objective reality, and that scientific progress consists in an increasing convergence on the truth about such a reality. But, if the history of science consists in repeated transitions between theories which refer to none of the same things, then it is impossible for progress to occur in the sense required by the scientific realist. For if later theories refer to none of the same things to which earlier theories referred, it is impossible for the transition between such theories to involve an increase of truths known about common items of a shared, objective reality.

For this reason, scientific realists have sought to defend Scheffler's appeal to reference against Kuhn's and Feyerabend's claim of referential variance. They have done so by drawing attention to the issue of the determination of reference. Kuhn's and Feyerabend's reference change response to Scheffler turns on the assumption that sense determines reference. But, as has been suggested by scientific realist advocates of a causal theory of reference, reference need not be determined by sense at all (e.g., Putnam 1975b, Ch. 11). Rather, reference is determined by means of various causal and other pragmatic relations which speakers enter with their environment in the course of linguistic interaction with the world. If this is right, then the reference of a term may be unaffected by variation of sense, and it may be possible to vindicate Scheffler's objection to incommensurability by setting the objection within the framework of a causal theory of reference.

Before explaining this response to the incommensurability thesis, let me briefly introduce the causal theory of reference. I will first discuss an example designed to show that reference is not determined by sense, and will then say something about the mechanism of reference.

Hilary Putnam presents a science fiction example which is designed to elicit intuitions favouring the view that reference is not determined by descriptive content (Putnam, 1975b, 223-7). This is his well known example of Twin-earth. We are to imagine a planet which is in many ways just like the earth. Its sole distinguishing feature is that the liquid which flows in its rivers, fills its oceans and falls from the sky in the form of rain is not in fact the same liquid as is found on earth. While the liquid found on earth is H_2O, the liquid on Twin-earth is another chemical compound, which Putnam calls XYZ. Yet despite its chemical difference, this substance is unable to be distinguished by any observable features from the water we find here on earth.

On Twin-earth there are also people who speak English. In particular, English speakers on Twin-earth use the word 'water' just as we do, to refer to the liquid that runs in their rivers and streams, and fills their lakes and oceans. Moreover, Twin-earth speakers of English associate exactly the same descriptive content with their word 'water' as we do with our word 'water'. We both conceive of water as the clear liquid which quenches thirst, extinguishes fires, falls from the sky as rain, and fills our lakes and rivers. There is no detectable difference whatsoever between the sense of 'water' as we use the term and the sense of 'water' as they use the term.

Now, Putnam asks, does the word 'water' as used by Twin-earth English speakers refer to the same substance as does the word 'water' as used by English speakers here on earth? Putnam argues that it does not. Our use of the word 'water' refers to H_2O, and Twin-earthians' use of 'water' refers to XYZ. Thus, in spite of the fact that English speakers on earth and Twin-earth associate the same sense with the word 'water', they refer to different things. Consequently, terms with the same sense may refer to different things, and so the sense of a term does not determine its reference.

The reason, Putnam argues, is that there is a broadly contextual element involved in the determination of reference. In our use of the word, the word 'water' was introduced by English speakers here on earth to refer to the stuff that as a matter of fact flows in our rivers and streams, and fills our lakes and oceans. In particular, when the word was introduced it was applied to paradigmatic samples of such stuff (e.g., glasses of water, or babbling brooks, or falling rain). The operative referential intention in such cases was to refer to the substance, whatever it happens to be, which is the same kind of substance as the paradigmatic samples of water. In other words, the word 'water' was introduced to refer to stuff that is the same kind as standard cases of the stuff that we here on earth call water.

The moral of the story is that speakers of a language need to be in some kind of causal relation to a thing in order to be able to refer to it. Ordinarily it is being in such a causal relation to a thing which determines that we do indeed refer to that thing.

The core idea of how reference is determined by causal relations is found in Saul Kripke's idea of an initial baptism (cf. Kripke 1980, 96-7, 135ff.). For proper names, Kripke suggests a commonsense approach to how people are named. For example, when a child is named, a ceremony takes place in which the baby is given a name. You can imagine the parents, looking at the baby in a cradle, saying 'We name that child William'. Given the parents' intentions to call their child William, and given that the context determines which child they are talking about, the reference of the name is fixed at such an initial baptism.

As for use of the name by speakers not present at the baptism, Kripke suggests that there is a causal chain linking later use of the name, via the use of other speakers, with the initial baptism at which the name was introduced. Unlike the use of speakers present at the initial baptism, use of the name by later speakers not present for the baptism depends on their being appropriately linked by a causal chain of earlier uses of the name back to that ceremony.

On this account, neither the reference of those present at the baptism nor of those linked to the baptism by a causal chain is determined by description. Rather, in the case of the introducers of the name reference is determined by such things as their intention to refer to the baby, perceiving the baby, being present when the naming takes place, etc. In the case of later use of the name, reference is determined by the causal chain linking later use back to the initial baptism. Thus, in general it is causal connections with the referent, rather than some shared descriptive content or sense, which determines reference.

This account may seem plausible for naming babies. But babies are particular observable objects, individual things which we may point to and perceive. There will be little confusion when we name a baby by pointing at it and giving it a name. But what about entire kinds of things like water? What about unobservable entities which we are unable to perceive, like electrons and protons?

The causal account of proper names may be extended to observable natural kinds of things, such as water. The basic idea is that, in the case of observable natural kinds, a term is introduced in the presence of a sample of the natural kind. Thus, when the term 'water' is introduced, it is applied to a sample of water. For example, I point to a glass of water and say 'By 'water' I refer to that stuff'.

The operative intention in introducing a term for a natural kind in the presence of a sample is not to attach the word specifically to that particular instance of the natural kind. Rather, the intention is to introduce the term as referring to the entire kind of which the sample is an exemplary instance. Thus, by saying 'That stuff is water', pointing to a glass of water, one is introducing a term to refer to the substance of which the water in the glass is a sample, namely H_2O.

In this way it is possible to extend the initial baptism account of naming from proper names to natural kind terms for observables. But what about theoretical natural kind terms, such as 'electron'?

One approach suggested by Kripke and Putnam is to extend the account to theoretical terms by means of causal descriptions employed in the presence of observable phenomena. On such an account, we are to imagine that a scientist, in the presence of an observable phenomenon thought to be produced by some specific unobservable causal process described by theory, may introduce a term to refer to the unobservable causal process. In this way the observable phenomenon may be picked out pragmatically by means of ostension, a causal description is given (e.g., 'The cause of that phenomenon'), and the term is thereby applied to the unobserved cause of the observable phenomenon.

The Causal-Theoretic Reply to Incommensurability

Assuming the causal theory of reference, one might reply to the incommensurability thesis somewhat as follows. The meaning, in the sense of 'sense', of scientific terms may well vary in the course of theoretical change. However, it does not follow that reference must also vary as a result of such change of meaning. For reference is not determined by sense, but by causal chains which link the present use of terms with initial baptisms at which their reference was fixed. So reference does not vary with the changes of descriptive content which occur during theoretical change. Hence, reference is held constant across theoretical transitions, and theories may be compared by means of reference. Thus, there is no referential discontinuity, no incomparability of content, and no incommensurability.

The causal theory of reference undoubtedly yields a promising rebuttal to the incommensurability thesis. For, if reference is continuous through theory change, then it is possible for later theories to assert more truths than earlier theories about common objects of reference. However, a number of serious problems face the causal theory of reference in the current context, which raise doubts about the effectiveness of this rebuttal.

The first problem arises because the original version of the causal theory of reference eliminates the possibility of reference change altogether. According to Kripke, reference is established at an initial baptism, and subsequent use of a term traces back to the initial baptism. But if the reference of a term is fixed at such a baptism, then reference cannot change. But there appear to be cases in the history of science in which the reference of terms has changed. So the causal theory of reference must be modified to allow for the possibility of reference change; e.g., by allowing usage subsequent to initial baptism to affect reference. But this permits reference variance between theories, which was what the causal theory was supposed to avoid.[8]

There are difficulties with the causal theory's account of how reference is determined for observational natural kind terms. Kim Sterelny describes a voyage to Mars:

> Suppose I go to Mars and come across a catlike animal: I introduce the term 'schmat'. Schmats are animals bearing a certain relation to this paradigm local schmat I have just encountered. But what determines which relationship this is? For the schmat will be a member of many kinds. A non-exhaustive list would include: physical object, animate object, animate object of a certain biochemical kind, animate object with certain structural properties, schmats, schmats of a certain sex, schmats of a certain maturational state. (Sterelny, 1983, 121)

The problem is that the causal relations involved in ostensive introduction of the term do not specify which of the numerous kinds instantiated by paradigmatic schmats is the kind designated by 'schmat'. The problem is known as the '*qua* problem', since it is a problem of how to refer to an object *qua* member of the relevant kind. The *qua* problem shows that some descriptive apparatus is required in order to pick out the relevant kind, e.g., a verbal specification of whether 'schmat' is the name of a genus or a species.

A further problem for the causal theory involves the reference of theoretical terms. Suppose we introduce the term 'phlogiston' to refer to that, whatever it is, which causes that phenomenon, pointing to a fire. If oxygen is what causes fire, then 'phlogiston' refers to oxygen. But phlogiston does not exist, so that rather than mistakenly referring to oxygen, the term 'phlogiston' fails to refer to anything at all.[9] This suggests that there is a need to build into the causal description of the referent of a theoretical term a certain amount of descriptive apparatus besides the mere causal description 'whatever causes that phenomenon'. Plausible suggestions include description of the natural kind to which the supposed cause belongs and a description of the causal role fulfilled by the unobservable causal agency.

---

8 For the objection that the causal theory rules out reference change in science, see Fine (1975).

9 The point that on an unmodified causal theory, 'phlogiston' would refer to oxygen is made by Enç (1976, 267-8).

4.7 Causal Descriptivism, Post-Baptismal Use and Translation Failure

The problems just highlighted spell the end for a pure causal response to incommensurability. For if reference may change, and description plays a role in reference, no dismissal of referential discontinuity of the kind previously sketched may be upheld. However, the causal theory may be revised in light of these problems to permit a more nuanced referential response to the incommensurability thesis. In particular, a modified causal theory may be used to defend referential comparison, while retaining key semantic insights on which the incommensurability thesis is based.

In this section, I will briefly summarize the approach to incommensurability which I have developed at length in my (1991) and (1994, Ch. 3). The modified causal theory which I favour is a form of causal descriptivism, according to which the causal relations on which reference is based must be supplemented by description. In the case of introduction of a kind term by ostension of a sample, reference is determined by the causal relation of perception supplemented by a descriptive specification of the kind (e.g., species, genus) to which the sample belongs. In the case of reference to unobservable theoretical entities, the causal relation which determines reference must be specified by description. Thus, the reference of theoretical terms is determined by description of the causal mechanism, whereby the action of unobservable referents is thought to produce certain independently specified (e.g., by ostension) observable phenomena.

A causal descriptive model of reference determination resolves the problems of indeterminacy of ostension and reference failure of theoretical terms. But it leaves the issue of reference change untouched. To meet the latter problem, the causal theory must grant post-baptismal use a role in reference determination.

Rather than being inalterably fixed at an initial baptism, reference may shift as a result of subsequent use of a term. After a term is introduced, speakers continue to apply the term to various items to which they take it to refer. In this way, to borrow a phrase from Michael Devitt, terms become *multiply grounded* in their referents (Devitt, 1981, 56-7, 191-5).[10] Reference change may occur due to shift in the *pattern of groundings* which a term has in its referents. For example, a term may be introduced to refer to a given kind, yet subsequently be applied to members of another kind. Because of the shift in the items in which it is grounded, the term comes to refer to the kind to which it was subsequently applied, rather than the kind for which it was introduced.

A further implication of the role of post-baptismal use in reference determination is that a term may be associated with multiple ways of fixing reference. As has been shown by Philip Kitcher, different tokens of a scientific term-type may have their reference fixed in different ways (e.g., 1978, 535). For example, the reference of some tokens of a term may be fixed by ostension of samples of a given substance,

10 By 'grounding', Devitt means perception of an object by virtue of which a name designates its bearer (1981, 278). For present purposes, grounding may be understood more generally as a relation between term and object(s) by means of which the term refers to the object(s).

while the reference of other tokens may be fixed by a description which specifies the role played by the substance in producing certain effects. Such diversity of reference determination may seem unified from the point of view of scientists, for whom the different means of fixing reference are merely alternative ways of picking out the same thing.

A modified causal theory of the kind just outlined serves as the basis for my approach to incommensurability. I defend a version of the referential response to the incommensurability thesis, roughly along the lines of Scheffler (1967). More specifically, I argue that the content of theories may be compared by means of shared reference, despite translation failure due to semantic variance. Suitably modified, therefore, the causal theory of reference has the resources both to reveal what is correct about the incommensurability thesis and to remove it as a threat to rational theory choice and scientific progress.

My argument for translation failure assumes that the way in which the reference of a term is determined is a semantic property which must be preserved in translation. There are two principal ways in which expressions employed by one theory may fail to be translatable into the vocabulary of another. First, it may not be possible to determine reference within one theory in a manner permitted by the other. For example, the causal role description required to fix reference for a theoretical term may be incompatible with the basic laws of the theory. Second, it may not be possible to determine reference within a theory in the same set of ways as in another. For example, within one theory the reference of a term may be fixed both by ostension and by theoretical description, while from the point of view of the other theory the two reference-fixing procedures fail to pick out the same kind. In either of these two ways, translation fails because the way reference is determined in one theory is unavailable in, or is precluded by, the other. (For details of how ontological and causal-explanatory commitments impose limits on reference determination, see my (1991, 229-34) and (1994, 81-95).)

Thus, it may be argued on the basis of a modified causal theory that translation fails between theories due to limits on reference determination in the context of a theory. Such translation failure is entirely consistent with comparison of the content of theories, which may proceed by means of shared reference. The question remains, however, of the extent to which such comparison is assured, given that reference may vary subsequent to initial baptism.

The answer, in short, is that while reference may vary with theory, it need not do so in wholesale fashion. On my view, reference changes in a piecemeal manner, dependent on the facts relating to the use of a given term. As such, reference is not necessarily subject to variation with overall shift in the way theories describe the entities which populate a common domain of application. Rather, it may remain stable through major alteration of the descriptive content which speakers associate with terms.

While reference may change through shift in the pattern of groundings of a term, this does not mean that all post-baptismal use affects reference. Rather, speakers typically acquire reference via a chain of communication from earlier use, despite reference not being inalterably fixed at the original introduction of a term. Hence, reference may remain stable even though later speakers do not associate the same descriptions with terms as earlier speakers. Thus, while reference may vary due to post-baptismal shift in the pattern of groundings, in a great many cases reference is determined by deference to earlier use.

In any event, the threat of reference change due to variation in descriptions which determine reference is much reduced on the present account. On a classic description theory of reference, substantial variation in the concepts employed by successive theories may result in radical discontinuity of reference, since the descriptions associated with the terms used by theories may not be satisfied by the same things. By contrast, description plays less of a role on my account. In the case of terms introduced by ostension to refer to observable natural kinds, the associated descriptive content is minimal (e.g., a categorical expression such as 'liquid' or 'species'). Thus, the reference of such terms may withstand considerable variation of associated descriptive content in the transition between theories. As for theoretical terms, whose reference is determined by description of causal role, significant scope exists for variation in descriptive content without change of reference, since the latter may vary without affecting the former.

To sum up, on the view I propose, translation may fail due to limits on reference determination within theories, yet the content of theories is comparable by means of reference. This approach allows for significant conceptual change in the transition between theories, while enabling the choice between such theories to be made on a rational basis. Given the possibility of common reference, moreover, it is possible for theories to advance on the truth about a shared domain of entities, in the manner suggested by scientific realism.

4.8 Taxonomic Incommensurability

Apart from allowing rational theory choice, the chief attraction of the present approach is that it permits analysis of the semantic issues relating to incommensurability within a realist framework. For it reveals how theories, which are untranslatable due to conceptual variance, may nonetheless entail comparable consequences about a shared, mind-independent reality. In the final two sections of this chapter, I will discuss two developments which raise questions about the realist commitments of this approach.

The first of these involves the formulation in Kuhn's later writings of a new, *taxonomic* version of the incommensurability thesis. According to Kuhn's later view, scientific revolutions are characterized by changes in the taxonomic schemes by means of which theories classify the entities in their domains of application (2000, 29-30). Such changes include redistribution of members among existing taxonomic categories, modification of criteria of category membership, and introduction of new categories. At the semantic level, taxonomic change gives rise to change in the meaning of preserved vocabulary, which in some cases involves change of reference. In the case of new categories, it may also result in introduction of new vocabulary which differs semantically from previous vocabulary.

According to Kuhn, the taxonomic scheme of a theory is represented linguistically by a *lexicon*, which is a structured vocabulary of natural kind terms (2000, 52).

Kuhn argues that terms from one lexicon fail to be translatable into another due to a restriction on relations between natural kinds, which derives from what he calls the *no-overlap principle* (2000, 92, 232). According to the no-overlap principle, members of one natural kind may only be members of another kind if one of the kinds is contained in the other as species is contained in genus (see Kuhn, 2000, 92). Thus, a term cannot be translated from one lexicon into another if its extension includes items which belong to distinct kinds within the rival taxonomy, since that would result in violation of the no-overlap principle (2000, 232).[11]

Translation failure due to inability to transfer a kind from one taxonomy to another is entirely consonant with the reference-based approach outlined in the previous section.[12] However, Kuhn sought to base a number of anti-realist claims on taxonomic incommensurability. I will sketch these claims, and then point out why such anti-realism does not follow from translation failure between lexicons.

Where scientific realists tend to see scientific progress in terms of convergence on truth, Kuhn was critical of both the idea of convergent progress and the realist view of truth. That science is not convergent, he took to follow directly from incommensurability (cf. 2000, 243-4). Speaking of propositions from Aristotelian physics, Kuhn wrote that:

> Using our conceptual lexicon, these Aristotelian propositions cannot be expressed – they are simply ineffable – and we are barred by the no-overlap principle from access to the concepts required to express them. It follows that no shared metric is available to compare our assertions about force and motion with Aristotle's and thus to provide a basis for a claim that ours (or, for that matter, his) are closer to the truth. (2000, 244)

Thus, according to Kuhn, science does not converge on truth because claims from rival theories cannot be expressed in a common lexicon, with the result that they cannot be compared for closeness to truth.

Kuhn's objection, though, was not simply to the idea of advance on truth, but to the very idea of truth as correspondence to reality. For he says that 'the subject of truth claims cannot be a relation between beliefs and a putatively mind-independent or "external" world' (2000, 243). His objection to the correspondence theory seems to turn on the relation between lexicon and reality:

---

11 How no-overlap leads to translation failure may be illustrated by one of Kuhn's own examples. Apart from planets other than the earth which are classified as planets by Copernican astronomy, the Ptolemaic kind *planet* includes the sun and the moon. Thus, to translate the Ptolemaic term 'planet' into the Copernican lexicon would require introduction into the Copernican taxonomy of a kind which contains items belonging to three distinct Copernican kinds, viz., *planet*, *sun* and *satellite*. But such a kind cannot be introduced into the Copernican scheme, since it treats entities which behave according to different laws as belonging to a uniform natural kind.

12 My requirement that reference determination be preserved in translation ensures that reference to kind is preserved in translation. This in turn insures that no kind term may be translated into a taxonomic scheme which precludes the relevant kind (for details, see my 1994, 96-7).

> ... lexicons are not ... the sorts of things that can be true or false. A lexicon or lexical structure is the long-term product of tribal experience in the natural and social worlds, but its logical status, like that of word-meanings in general, is that of convention. Each lexicon makes possible a corresponding form of life within which the truth or falsity of propositions may be both claimed and rationally justified, but the justification of lexicons or of lexical change can only be pragmatic. With the Aristotelian lexicon in place it does make sense to speak of the truth or falsity of Aristotelian assertions in which terms like 'force' or 'void' play an essential role, but the truth values arrived at need have no bearing on the truth or falsity of apparently similar assertions made with the Newtonian lexicon. (2000, 244)

Thus, Kuhn was prepared to grant a role to truth within the context of a given lexicon. But it is a limited role. Neither is the truth of claims made in one lexicon relevant to that of claims made in another, nor may the concept of truth be applied to a lexicon itself. On the concept of truth which emerges, truth is internal to lexicon in the sense that a claim may be true in a lexicon without corresponding to reality.

But Kuhn seems simply to have been mistaken in taking these anti-realist claims about truth and convergence to follow either from taxonomic incommensurability or from the nature of lexicons. It is a mistake to suppose that inability to translate between lexicons entails either that theories cannot converge on truth or that they may not be judged to do so. For, even if two theories are not fully intertranslatable, it remains possible for terms used by such theories to refer to some of the same things. To the extent that there is co-reference, such theories may assert true or false claims about a common domain of entities. Indeed, one theory may even assert a greater number of truths than the other about such entities.

Nor does untranslatability preclude comparative judgement of truth. For a lexicon is a special vocabulary, embedded in a background natural language. Given this, it is possible to employ a fragment of the natural language as metalanguage with respect to alternative lexicons, which may be treated as object languages. Within such a metalanguage one may assert the truth or falsity of object-linguistic sentences from alternative lexicons. Therefore, it is at least in principle possible to arrive at a judgement of comparative closeness to truth of theories expressed in non-intertranslatable lexicons.

As for the realist idea of truth, Kuhn appears to reject the idea of correspondence between theory and reality on the basis of the conventional status of lexicons. He is right, of course, about the conventionality of lexicons, since the words which make up a lexicon are conventionally associated with their meanings. But nothing follows from this about the nature of truth. Statements may be true because what they say corresponds to the way things stand in reality, regardless of the conventional status of lexicons.

In any event, to assume that the status of a lexicon, in relation to truth, is entirely a matter of convention is to overlook the factual nature of claims which may be made on the basis of a lexicon. For it is possible to use the terms of a given lexicon to formulate substantive claims about the world regarding what kinds of things there are, as well as the properties that such things possess. To assume that such claims are lacking in truth-value is to assume that there is no fact of the matter relating to the structure or constitution of reality. But such an extreme metaphysical thesis is certainly not entailed by the conventionality of lexicons.

My point, though, is not just that the anti-realist claims specifically based by Kuhn on taxonomic incommensurability are unfounded. Rather, at a more general level, the thesis of taxonomic incommensurability poses no threat to scientific realism. The phenomenon of conceptual change between theories is consistent with a commitment to the reality of theoretical entities, which is characteristic of scientific realism. Moreover, variation of the taxonomic schemes whereby theories classify the world does not by itself have any bearing on metaphysical matters. The translation failure between theories, to which it gives rise, is a semantic relation between alternative theories of the same world.[13]

4.9 Meta-incommensurability?

In the last section, I mentioned fit with realism as an attraction of my approach. But not all readers may take such fit as a positive feature. Some will see it as a drawback. I will conclude this chapter by commenting on a suggestion that realism is in fact positively inimical to understanding incommensurability.

The suggestion is due to Hoyningen-Huene, Oberheim and Andersen (1996) – HHOA, for short – in a critical study of my (1994) book. The main impetus behind the suggestion derives from the proposal by Paul Hoyningen-Huene (1993) of a novel interpretation of Kuhn's philosophy of science, which presents the latter within a neo-Kantian anti-realist framework. On this reading of Kuhn, the incommensurability thesis involves anti-realist assumptions, which problematize a realist critique of incommensurability.

Hoyningen-Huene takes seriously Kuhn's talk of the world changing with change of paradigm. To make sense of such talk, he proposes a distinction between the variant *phenomenal world* of the scientist and the invariant *world-in-itself* (1993, 33-5). The phenomenal world is a world of appearances, jointly constituted by perceptual input from external reality and the conceptual contribution of an epistemic subject. By contrast, the world-in-itself is a fixed, independent reality, not subject to the influence of human conceptual activity. Thus, in the transition between paradigms it is the phenomenal world which varies, while the world-in-itself remains constant.

The metaphysical divide between phenomenal and real corresponds to an epistemological divide between the knowable and the unknowable, reminiscent of Kant. On Hoyningen-Huene's interpretation of Kuhn, scientific knowledge is knowledge of a phenomenal world, since the world-in-itself is unknowable and epistemic access is limited to the phenomena. However, the chief novelty of such an interpretation lies not in its Kantian underpinning, but in the analysis it proposes of Kuhn's theory of the constitution of phenomenal worlds.

Drawing mainly on Kuhn's account of empirical concept-acquisition in 'Second Thoughts on Paradigms' (1977, Ch. 12), Hoyningen-Huene argues that subjects enter

13 For more detailed discussion of the matters dealt with in this section, see my (1997, Ch. 4).

the phenomenal world of a scientific community as they acquire the community's conceptual system (1993, 70-111). A major part is played in concept-acquisition by ostension, whereby a subject is taught a set of similarity relations by direct exposure to members and non-members of a given similarity class or kind. For concepts at the more theoretical end of the spectrum, exposure to sets of similar problem situations takes the place of ostension. The network of similarity relations acquired by these processes is what provides the conceptual ordering of the phenomenal world, which the subject acquires on initiation into a given scientific community.

Such a reading of Kuhn suggests that incommensurability is also to be taken in anti-realist vein, as a relation between theories about different phenomenal worlds rather than objective reality. That incommensurability is an essentially anti-realist idea is the basic assumption of HHOA in their (1996). HHOA write, for example, that for Kuhn and Feyerabend 'incommensurability is one form of expressing a critical attitude toward naive realism' (1996, 133). 'Realism', they say, 'is just the issue of the dispute' (1996, 135).

By contrast, my approach makes 'realist assumptions that lead [me] to misconstrue Feyerabend and Kuhn's intentions in establishing the incommensurability thesis' (1996, 131). This results in misunderstanding, perhaps due to a '*meta-incommensurability* between the realist and the non-realist' (1996, 138), which involves meaning change, communication difficulty and circular argument (1996, 139-40). Thus, HHOA suggest, debate about incommensurability is vitiated by a meta-level incommensurability arising from 'the different metaphysical commitments realists and non-realists bring to the debate' (1996, 141).

The claim of meta-level semantic variance is not, of course, implausible, as may be illustrated by comparison of my use of the term 'realism' with that of HHOA. By 'realism' I mean *scientific realism*, which is a doctrine typically committed to correspondence truth, reality of theoretical entities and advance on truth about mind-independent reality. In contrast, for HHOA 'realism' refers to the view that scientists' epistemic access extends to objective reality, rather than being limited to a changing phenomenal world, as it is for what they call 'non-realism' (cf. 1996, 139). While such discrepancy of use may create temporary confusion, the semantic adjustment required is routine: adopting a metalinguistic mode of discourse, one either explicates the meaning of a mentioned term or else specifies its reference, as I have just done. In short, I can see no basis in the semantic variation of philosophical terms for a crippling meta-level incommensurability between realist and non-realist.

HHOA make much of the supposed intentions of Kuhn and Feyerabend. The incommensurability thesis was 'originally intended', they say, 'as a challenge to realism' (1996, 138). Now, it would undoubtedly be an important *historical* discovery, if it were shown that Kuhn and Feyerabend meant all along for incommensurability to be understood in the context of an anti-realist position of the kind described by Hoyningen-Huene.[14] But such a discovery would not detract in the

---

14 While I do not propose to challenge the claim about Kuhn and Feyerabend's intentions in detail, I am quite doubtful of this claim. With respect to Kuhn, suffice to say that while Hoyningen-Huene's neo-Kantian interpretation makes sense of Kuhn's early talk of world-change, it hardly follows that Kuhn *intended* such talk to have exactly that interpretation. As

least from the *philosophical* importance of the semantic incommensurability thesis standardly discussed in the literature on the topic. For, as we have seen, semantic incommensurability spells trouble for scientific rationality and progress – to say nothing of the problems it raises for traditional views of methodology.

Setting aside the question of authorial intent, the claim that incommensurable theories relate to different phenomenal worlds is a distinct thesis from the thesis of semantic incommensurability. Let us call it the Φ-incommensurability thesis. I will close this discussion by briefly outlining why, in my view, the Φ-incommensurability thesis fares no better than the more standard semantic version of the thesis.

In the first place, while Hoyningen-Huene (1993) proposes a very plausible interpretation of Kuhn's metaphysics, the plausibility of the position thereby ascribed to Kuhn is itself somewhat doubtful. Indeed, the claim that there is a world-in-itself, which is itself unknowable, appears to be fundamentally incoherent. For how can it coherently be said *of* an unknowable reality that *it* is unknowable?

In order to know that such a reality cannot be known, it must be possible to enter into *some* sort of epistemic relation with it, namely, the epistemic relation of knowing that it cannot be known. At the very least, this suggests that it is possible for a cognitive agent to enter into an intentional or referential relation with reality itself, on the basis of which it is possible to bear an epistemic relation to it. This, in turn, raises the question of why more extensive knowledge of reality itself should be precluded, given that it is possible to enter *some* epistemic relation to it.[15]

The problem may be brought into focus by consideration of Kuhnian anomalies. Failure of theory-data fit, characteristic of anomaly, requires that the world-in-itself be capable of impact on the phenomenal world of the scientist. To account for the existence of anomalies, Hoyningen-Huene speaks of the 'proprietary resistance of the world-in-itself' (1993, 270). 'The world-in-itself', he writes, 'offers resistance, resistance which makes it impossible to impose just any network of similarity relations' (1993, 269). Such resistance 'reveals itself when ... significant anomalies appear'; 'resistance of the world-in-itself must be a participant in the production of significant anomalies' (1993, 270).

The trouble with such talk of resistance is that it is irreducibly *causal*.[16] For in order for the world-in-itself to resist, and thereby induce change in phenomenal

---

for Feyerabend, the claim about original intentions will prove difficult to reconcile with the sustained polemic against positivism in the name of realism, which Feyerabend conducted in the late 1950s and early 1960s. More generally, the claim about intentions misidentifies the opposition. At the time of the introduction of the incommensurability thesis in the early 1960s, logical empiricism was the target, not scientific realism.

15 An analogous problem arises for Kuhn, who writes that: 'Underlying all these processes of differentiation and change, there must, of course, be something permanent, fixed, and stable. But, like Kant's *Ding an sich*, it is ineffable, undescribable, undiscussible. Located outside of space and time, this Kantian source of stability is the whole from which have been fabricated both creatures and their niches, both the "internal" and the "external" worlds' (2000, 104). Yet for something that cannot be described, Kuhn manages to say rather a lot about it.

16 Indeed, this is evident from the causal idiom which Hoyningen-Huene adopts in discussing resistance: e.g., resistance 'makes itself felt', and it is 'a participant in the production

world, it must interact causally with the phenomenal world. Anomaly is *caused* by unpredicted behaviour of mind-independent entities, just as recalcitrant experience is *caused* by impact of such entities on our sensory receptors. But given the existence of causal relations between world-in-itself and phenomenal world, there is no basis on which to deny that there may be causally grounded referential relations linking scientists' linguistic tokenings to mind-independent objects. In short, Hoyningen-Huene would seem to have no reason to deny that terms of semantically variant theories may refer to common entities in the world-in-itself.

Hoyningen-Huene is not, in any case, opposed in principle to intertheoretic co-reference. Indeed, in his discussion of comparison of Φ-incommensurable theories, he explicitly allows for such co-reference (1993, 219-20). In some cases, such comparison proceeds on the basis of shared, semantically stable vocabulary:

> ... some of the empirical predictions of incommensurable theories can be compared *immediately*, namely those unaffected by the (merely local) incommensurability of the lexica. (1993, 219)

For more complex cases, Hoyningen-Huene follows Kitcher in admitting theory comparison on the basis of the co-reference of tokens of semantically variant term-types:

> ... there may be *particular situations* in which the referents of the new concepts may be identified by means of the concepts of the old lexicon. (1993, 220)

In either case, comparison of the content of theories proceeds by means of common reference. But if terms from Φ-incommensurable theories co-refer across phenomenal worlds, they must refer to the same items in the world-in-itself. Given this, Hoyningen-Huene would appear to be committed to shared reference by the terms of Φ-incommensurable theories to common elements of an objective reality.

Nor is there any reason for Hoyningen-Huene to demur. For why should he not allow co-reference to mind-independent objects across phenomenal worlds? After all, as Hoyningen-Huene notes, 'a phenomenal world is constituted both by the object-sided world-in-itself and by subject-sided moments' (1993, 36).[17] Given this, the elements which constitute the phenomenal worlds of Φ-incommensurable theories must include items drawn from the *same* domain of objects in the world-in-itself.[18]

---

of significant anomalies' (1993, 270); we even 'feel the effects' of 'concrete properties of the world-in-itself' (1993, 271). The point is made explicitly by Hoyningen-Huene (1993, 34), where he talks of the 'causal efficacy' of the world in 'codetermin[ing] a given phenomenal world'.

17 The terms 'object-sided' and 'subject-sided' are introduced by Hoyningen-Huene (1993, 33ff) to distinguish between that which is independent of human cognition (e.g., objective reality) and that which derives from the epistemic subject (e.g., concepts, ideas).

18 The strain inherent in Hoyningen-Huene's position is particularly evident at this point. In an attempt to explain how incommensurable theories differ from theories which really are about different domains, he writes that: 'incommensurable theories, by contrast [with such theories], target *roughly the same* object domain, *as far as the world-in-itself is concerned*' (1993, 219). Such appeal to the world-in-itself to individuate a common object

For where such phenomenal worlds differ is not in their objective constituents, but in the conceptual ordering overlaid by theories on such constituents.

The upshot is that the phenomenal worlds of Φ-incommensurable theories contain phenomenal items, the constitutive parts of which include the same mind-independent elements of the world-in-itself. But, given that phenomena in different phenomenal worlds may be constituted of the very same objective elements, scientists working in such different worlds may nevertheless refer to the same objective entities. For, while they employ terms which refer to items in their phenomenal world, they thereby refer to the objective entities which partly constitute those phenomenal items. As a result, there may be co-reference across Φ-incommensurable theories to common entities in the world-in-itself.

The possibility of co-reference across phenomenal worlds combines with the causal nature of resistance to yield an unexpected result. Given both co-reference and resistance, little substantive difference remains between Hoyningen-Huene's position and my own. Rather than speak of phenomenal worlds and an inaccessible world-in-itself, one may simply say that there is one world about which different folks believe different things. Far better to say, as I do, that the vocabulary of semantically variant theories may refer, by way of causal links, to the *same* mind-independent objects, though these *same* objects may be categorized in different ways by different theories.

---

domain illegitimately trades on facts about the world-in-itself which Hoyningen-Huene otherwise treats as inaccessible. Quick to correct the slip, however, he adds: 'Such talk of "object domains" obviously can't be taken literally, as this notion properly applies only to regions of the phenomenal world' (1993, 219, fn. 119). The problem is, of course, that there is no way to individuate a common domain for theories without appeal to a shared set of objects, such as may be provided by the world-in-itself.

Chapter 5

Induction and Natural Kinds

5.1 Introduction

In this chapter, I present a novel approach to the problem of induction. The approach derives from the basic realist intuition that it is the objective structure of the world that underpins the success of our epistemic practices. Crudely put, it is because the world is the way it is that we are right to use induction to form beliefs about the future. I seek, in short, to rehabilitate the principle of the uniformity of nature, and to employ the principle of uniformity to answer Hume's problem of induction.

The approach I propose is based on three distinct philosophical positions. The first position is the position of scientific realism, which holds that the aim of science, towards which a great deal of progress has been made, is to discover the truth about both observable and unobservable features of an objective, mind-independent reality. The second is scientific essentialism, the view that the objective reality investigated by science is populated by mind-independent natural kinds of things, which are characterized as such by the fundamental, intrinsic causal powers which they possess. The third position is epistemic naturalism, which treats epistemological questions about normative justification as broadly empirical questions about how best to pursue inquiry into the objective, natural world which we inhabit.

At the most general level, my strategy is to employ an *ontological* theory about the constituents of reality to solve an *epistemological* problem about the normative justification of inductive inference. In particular, I wish to appeal to the existence of natural kinds, whose members have inbuilt, causal powers, to explain why our use of inductive inference is epistemically justified.

In outline, my argument will run as follows. Induction is justified because there are, in reality, natural kinds of things. These natural kinds are characterized by sets of properties which all members of a natural kind possess essentially. Thus, when one makes a correct inductive inference about unobserved members of a natural kind, what makes it true that unobserved members of the kind have the properties we predict them to have is that they are members of a natural kind all of whose members possess those properties essentially.

5.2 The Problem of Induction

I understand the problem of induction in a straightforward fashion. The problem is how to justify induction. Two options at first suggest themselves. Either induction may be justified on the basis of logic, or on the basis of experience.

If 'logic' means deductive logic, there can be no logical justification of induction, since inductive inference is deductively invalid. As for experience, if it is said that induction is justified because induction has worked in the past, then induction is employed to support induction, which is circular.

Here one might appeal to a principle of the uniformity of nature, which says that induction succeeds because nature is uniform. The reason induction successfully predicts the future is that, since nature is uniform, the future resembles the past. Because of future resemblance to the past, past regularities will continue in the future.

I think that such an appeal to the uniformity of nature is basically on the right track. The trouble is, of course, that appeal to the uniformity of nature seems itself to depend on induction. For, to argue that nature is uniform, one must infer from past uniformity to future uniformity, which is itself an inductive inference from past to future. Given this, appeal to the uniformity of nature proceeds in a circle, since it presupposes induction.

In this chapter I sketch a scientific essentialist response to this standard objection to the appeal to the uniformity of nature.

5.3 Kornblith's Epistemic Naturalism

The approach I propose is an extension of the approach taken by Hilary Kornblith in his book, *Inductive Inference and its Natural Ground* (1993a). Kornblith proposes an account of the success of inductive inference, which is based on a naturalized epistemology and a realist metaphysics of natural kinds. He distinguishes between two questions: 'What is the world that we may know it?', and 'What are we that we may know the world?' (1993a, 2). In answer to the first question, he argues that:

> ... natural kinds make inductive knowledge of the world possible because the clustering of properties characteristic of natural kinds makes inferences from the presence of some of these properties to the presence of others reliable. (1993a, 7)

As for what it is about us that enables us to know the world, Kornblith argues that evolution equips the human mind with conceptual structures and inferential strategies which make it sensitive to the natural kind structure of the world. Thus, there is what he describes as a 'dovetail' fit between mind and world, in virtue of which inductive knowledge is possible.

Kornblith appeals to the success of science to argue both that there is a fit between mind and world, and that natural kinds ensure the reliability of induction. I will consider only the second argument.

Kornblith favours an account of natural kinds, due to Richard Boyd (1991), according to which natural kinds are *homeostatic property clusters*. On this view, according to Kornblith,

> A natural kind is a cluster of properties which, when realized together in the same substance, work to maintain and reinforce each other, even in the face of changes in the environment. (1993a, 35)

Kornblith claims that because the properties which define natural kinds 'reside in homeostatic relationships, we may reliably infer the presence of some of these properties from the presence of others' (1993a, 36). Moreover, were it not for the fact that only certain groups of properties are capable of sustaining such homeostatic relationships, 'the presence of any set of properties would be fully compatible with the presence of any other' (1993a, 36).

According to Kornblith, the best explanation of the success of induction in science is that there exist real natural kinds whose existence ensures the reliability of induction:

> If the scientific categories of mature sciences did not correspond, at least approximately, to real kinds in nature, but instead merely grouped objects together on the basis of salient observable properties which somehow answer to our interests, it would be utterly miraculous that inductions using these scientific categories tend to issue in accurate predictions. Inductive inferences can only work, short of divine intervention, if there is something in nature binding together the properties which we use to identify kinds. Our inductive inferences in science have worked remarkably well, and, moreover, we have succeeded in identifying the ways in which the observable properties which draw kinds to our attention are bound together in nature. In light of these successes, we can hardly go on to doubt the existence of the very kinds which serve to explain how such successes were even possible. (1993a, 41-2)

It is noteworthy here that, instead of arguing for the reliability of induction, Kornblith takes such reliability as something given that is in need of explanation. His claim is that the best explanation of the reliability of induction is that there exist real kinds in nature, which our inductive inferences latch on to.

It is evident, then, that Kornblith does not seek to justify induction in the sense of giving a non-circular basis for expecting induction to continue to be reliable in the future. Rather, he assumes that induction is reliable in order to argue that natural kinds must exist, since their existence is the best explanation of its reliability. In other words, he argues for the existence of natural kinds on the basis of the reliability of induction, rather than arguing for the rationality of induction, given their existence.

In arguing this way, Kornblith's project shows its true Quinean colours. For, in simply assuming the reliability of induction, and thereby failing to confront inductive scepticism, Kornblith's approach is informed both by what he at one point describes as a 'robust antiskepticism' (1993a, 4-5) and by an approach to epistemological problems which derives from Quinean naturalism. While I am broadly sympathetic both to anti-scepticism and epistemic naturalism, I wish to depart from Kornblith by confronting inductive scepticism somewhat more directly. In particular, I wish to argue against the inductive sceptic that it is the existence of real kinds in nature which justifies our use of induction.

5.4 Ellis's Scientific Essentialism

On the view I propose, we may employ the principle of the uniformity of nature to justify induction. But I do not understand the principle of uniformity in the usual way.

Customary formulations of the principle of uniformity involve claims to the effect that the future will resemble the past, or that things which have always occurred together previously will continue to do so. But rather than understand the uniformity of nature as any sort of blanket resemblance of past to future, I understand it in terms of the law-governed behaviour of objects.

More specifically, I take the view that the laws of nature are at base the inbuilt causal powers of things which belong to natural kinds. This view, which has been developed in detail by Brian Ellis, is the position of *scientific essentialism.* I will introduce the leading ideas of Ellis's essentialist theory of laws of nature, and then explain how this theory accounts for the uniformity which underlies induction.

Ellis presents his theory of laws of nature in opposition to what he sees as the dominant metaphysic of contemporary Anglo-American philosophy, which he characterizes as both mechanist and Humean. According to this metaphysical view, matter is essentially passive or inert. The behaviour of material objects is, however, governed by laws of nature, which constitute nothing more than empirical regularities which hold universally. These laws of nature are contingent regularities, which are entirely distinct from the nature or intrinsic properties of the material objects themselves. Hence, the very same set of objects which exist in our world might exist completely unchanged in some other possible world that is governed by an entirely different set of natural laws.

By contrast with the mechanist or Humean metaphysic, Ellis argues that the fundamental properties of material objects are not passive but active. They have, he says, 'the nature of powers, capacities and propensities' (1999, 19; cf. 2001, 106). Such powers, capacities and propensities cannot be reduced to more basic inactive or categorical properties of things. Rather, they constitute the irreducible dispositions of things to behave in certain ways under given circumstances.

The view that the behaviour of material objects is due to irreducible causal powers leads Ellis to reject the Humean account of laws as mere empirical regularities. For, given that objects possess real causal powers, the laws of nature must be something more than the mere regular behaviour of inert matter. Rather, the laws of nature, for Ellis, are descriptions of the behavioural patterns of things, which are made true by the possession by those things of real, inbuilt causal powers.

Given that the most basic properties of things are irreducible causal powers, it follows that the laws of nature are necessarily true. For if material things may not lose their fundamental properties without ceasing to be what they are, the possession of those properties is crucial to their identity. But since laws of nature depend on the basic causal powers of things, in any world in which the things exist the laws describing the causal powers must be true. For they would not be those things if they did not possess those powers.

As for natural kinds, Ellis argues that the fundamental kinds of things are characterized as such by the intrinsic causal powers of things belonging to those kinds, which constitute the Lockean real essences of such kinds (1999, 22; 2001, 31). These powers are, at base, the irreducible dispositional properties of the individual members of natural kinds, and it is the behavioural manifestations of these dispositions which are described by the laws of nature which govern causal processes and interactions (cf. 1999, 21-2; 2001, 127-9). Because it is the intrinsic causal powers or dispositional properties which constitute the real essences of natural kinds, possession of such powers or properties is crucial to the identity of natural kinds. In particular, if a thing is to be a member of a given natural kind, then, necessarily, it must possess the set of intrinsic powers and dispositions which are essential to that kind. If it does not do so, then it cannot be a member of that kind.

5.5 An Essentialist Response to Hume

As we saw before, Kornblith argues that the best explanation of the success of scientific induction is the existence of natural kinds, which possess homeostatic property clusters, the co-occurrence of which assures the reliability of induction. This argument takes the success of induction as a given fact, and seeks to provide an explanation for this success. As such, Kornblith fails to address Hume's sceptical challenge, which is to show that we are rationally justified in expecting induction to succeed in the future.

As hinted previously, I wish to argue that induction is justified because nature is uniform. I do not understand the principle of the uniformity of nature as any sort of blanket assertion that the future resembles the past. Rather, I see it as grounded in the properties of individual substances. More specifically, nature is uniform in the sense that it contains natural kinds, all of whose members possess a common set of essential properties.

We may think of the claim that nature is uniform as the claim that the world is governed by laws of nature. This is because the essential properties of things are, in fact, the fundamental causal capacities of members of natural kinds. So, when we discover that observed phenomena are governed by laws of nature, this is because we are discovering natural uniformities which are grounded in the basic causal powers of things which belong to natural kinds.

My response to Hume, then, is that we are rational to employ induction when we form our beliefs about the future because nature is, in fact, uniform. It is uniform in the sense that the fundamental kinds of things which exist are natural kinds of things, which possess essential sets of properties. Because all members of a kind possess the same essential properties, unobserved members of a kind will possess the same properties as members of the kind which have already been observed.

This is why, when we infer that an unobserved object will have a property which observed objects of the same kind have, we turn out to be right. For having such a property is just part of what it is to be an object of the same kind as the other objects. Thus, what makes it rational to make inductive predictions about objects which belong to kinds is simply that it is part of the nature of objects of a given kind to have certain properties. So, as Kornblith rightly argues, it is nature that grounds inductive inference.

5.6 Objections and Replies

I do not claim that the approach to the problem of induction that I have proposed here constitutes the final solution of the problem. However, I do suggest that it is a promising approach to the problem which sits well with a number of influential contemporary philosophical tendencies. In particular, the approach should appeal to philosophers who are otherwise attracted to the combination of scientific realism, essentialism and epistemic naturalism, on which the current approach is based. As presented here, the approach remains simplified and incomplete. I will conclude by addressing a number of objections which might be raised against the position that I have outlined here.

Objection one: The claim that induction is justified by the existence of natural kinds must ultimately run in a circle. For the claim rests on the doctrine of scientific essentialism, which is itself based on an inference to the best explanation. More specifically, scientific essentialism is based on the argument that the existence of natural kinds with essential properties is the best explanation of the success of science. But inference to the best explanation is itself a form of inductive inference. Thus, in the end, the appeal to natural kinds to justify induction must use induction to justify induction.

Reply: Let us grant that inference to the best explanation is a form of induction, at least for the sake of argument.[1] Even so, it still does not follow that the proposed justification of induction is circular. For the inference to the best explanation, on which scientific essentialism is based, is a quite distinct argument from the inductive inferences which the existence of natural kinds serves to ground.

This point may be best seen if we make the simplifying assumption that the sort of inductive inference at issue is low-level enumerative induction about the properties of observable kinds of things. Such an inference has the form:

> All observed As have been Bs.
> Therefore all As are Bs.

Such an enumerative inference proceeds from a premise about the features of observed As to a conclusion about the features of unobserved Bs. As such, it contrasts sharply with the argument for scientific essentialism, which might be cast as follows:

> Science is successful.
>
> The existence of natural kinds is the best explanation of the success of science.
> Therefore, there are natural kinds.

An argument of this kind is an argument for the existence of natural kinds which proceeds from the claim that their existence is the best explanation of the success of science. Thus, at the very least, it may be said that the present defence of induction

1 Why only for the sake of argument? Because it is not, in fact, clear that inference to the best explanation is best construed as a form of inductive inference. One option, due to Musgrave (1988, 238; 1999, 234) is to construe inference to the best explanation as a deductive enthymeme which tacitly rests on the epistemological principle that it is reasonable to tentatively accept the best explanation as true (see the discussion in 8.3). Another option, suggested by Armstrong (1995, 48ff), is to construe inductive inference itself as a form of inference to the best explanation (cf. Armstrong, 1983, 52-3). (See also Harman 1965.)

is not circular in the sense of using enumerative induction to justify enumerative induction.

Objection two: The reply to the first objection only has the effect of prolonging an inevitable admission of defeat. For even if it is enumerative induction that is defended on the basis of an inference to the best explanation of the success of science, it is still ultimately induction that is being used to justify induction. And *that* is circular.

Reply: Here appeal may be made to a point that has been made in the context of inductive arguments on behalf of the reliability of induction. David Papineau, for example, points out that the circularity involved in using induction to argue that induction is reliable is rule circularity, rather than premise circularity (Papineau, 1992). The conclusion of the inductive defence of induction (viz., that induction is reliable) does not occur as a premise in the argument, though the rule (viz., induction) defended by the argument is employed in the defence of that very rule. While premise circular arguments are viciously circular, it is difficult to see how to avoid use of rule circular arguments at this level of generality. Indeed, it is not clear what is objectionable about them. After all, similar considerations reveal deductive arguments in support of deduction to be circular in the sense of rule-circularity.

Objection three: The admission of rule circularity leads back to the problem of induction, which the present approach is meant to address. For the present approach will carry no weight against the sceptic unless the sceptic is prepared to grant the force of inference to best explanation. But, given the inductive character of inference to best explanation (assumed in reply to Objection One), this is to assume precisely what the sceptic challenges, namely, the legitimacy of induction. (For this objection, see Lange, 2004, 199.)

Reply: From the point of view of the epistemic naturalist, the sceptic sets the epistemological bar too high (see Section 1.4). A reply that satisfies the sceptic is not therefore to be expected on behalf of the present approach. The approach is a naturalistic one which treats epistemic normativity in an empirical fashion. It takes seriously the epistemic claims of empirical science, and adopts the best explanation of the success of such science. In its use of inference to best explanation, the approach employs a form of inference that is routinely employed in both commonsense and scientific contexts. It therefore appeals to no higher tribunal than the epistemic standards employed in common sense and scientific inquiry. While this may not satisfy the sceptic, the naturalist's point is just that the sceptic's appeal to standards over and above those of common sense and science is itself to be rejected as inappropriate.

Objection four: The approach rests on a fundamental mistake. It attempts to solve an epistemological problem on the basis of ontological assumptions. But ontological assumptions may not be employed to solve an epistemological problem, since epistemological problems are prior to ontological ones. No assumptions may justifiably be made about the nature of reality until it has been shown that we know both that there is an external world and that we are able to have knowledge about the nature of such a world. Hence, to appeal to ontological considerations to resolve an epistemological problem is to invert the proper order of philosophical argumentation.

Reply: It is, in the first place, implausible to suppose that epistemology may proceed entirely in the absence of ontological assumptions. Epistemological reflection on the nature of knowledge must include reflection on what knowledge is in general. It must include reflection on the nature of knowers or epistemic agents, as well as on the kinds of things which may be the object of knowledge. But, given that appeal must be made to both knowers and things that may be known, it would seem that no epistemology can proceed in the complete absence of ontological assumptions.

Admittedly, the present account of the justification of inductive inference does deploy a rather rich array of ontological assumptions about such things as natural kinds and essential properties. One might think that such assumptions go well beyond what is needed to address an epistemological problem. However, it is not clear that there is any need to reduce ontological assumptions merely for the sake of minimizing such assumptions. If the correct metaphysical picture is one which includes natural kinds and essential properties, and the existence of such kinds and properties has implications for inductive inference, then there is no reason not to include such assumptions when one turns to matters epistemological.

Objection five: Induction cannot be grounded in natural kinds, for there are sound inductive inferences about artefacts and other non-natural kinds. One may infer inductively about cars, for example. Thus, in the past, moving into the path of a rapidly moving car has been dangerous, hence in the future moving into the path of a rapidly moving car will also be dangerous.

Reply: It is true that good inductive inferences are not restricted to natural kinds. One may make good inductive inferences about artefacts such as cars. What is not clear is whether the artefactual nature of the items concerned is relevant to the success of inductions about such things. What is dangerous about moving into the path of a moving car is not just that one may be struck by the car, but that one might get struck by an object with a large mass moving at a high velocity.[2] The risks involved are not dissimilar to those involved in being run down by a charging buffalo, or struck by a boulder rolling down a hillside or a falling meteorite. Thus, it may very well be the case that one is only able to make reliable inductive inferences about artefacts and other non-natural kinds, to the extent that such inferences turn on facts about them which obtain in virtue of their being members of natural kinds.[3]

Objection six: The account only applies to the essential properties of natural kinds. Yet not all inductive inferences concern the essential properties of kinds. There might be inductive inferences which range over the accidental properties of kinds of things. Hence, the account mistakenly assumes that inductive inference is restricted to the essential properties of things.

---

2 Anjan Chakravartty objects that this appeals to a "tailor made" kind, namely the kind "heavy, fast objects" (2007,169). But I would be happy instead to countenance mass, and perhaps velocity, as natural kind properties, rather than hypostatizing such a made-up kind.

3 T.E. Wilkerson makes the point well: '... because there are no very specific real essences that make rubbish rubbish, or tables tables, I cannot even in principle make sound inductive projections about rubbish as such or tables as such' (Wilkerson, 1995, 32).

Reply: I conjecture that a crucial distinction between good and bad induction hinges on the distinction between essential and accidental properties. For the most part, good inductive inferences project essential properties, whereas bad ones project accidental properties. One might perhaps distinguish three cases of possible relevance here: (i) inductive inferences which range over essential properties, (ii) inductive inferences which range over accidental properties, (iii) inductive inferences which range over accidental properties that depend on essential properties. Inductions of type (i) tend to succeed, those of type (ii) tend to fail, and those of type (iii) may succeed if the relevant link between essential and accidental properties obtains.

Objection seven: The account is, perhaps, plausible for induction in the physical sciences, where the fundamental particles constitute natural kinds. But there are no natural kinds in biology. Yet surely there is sound inductive inference in the life sciences.[4]

Reply: Any account of scientific induction must apply to induction in the life sciences, as well as in the physical sciences. But the present account need not address this issue. What the objection really does is to raise the issue of reduction. If biological kinds can be reduced to physical kinds, then inductive inferences which range over biological kinds will hold because biological kinds reduce to physical kinds. But if biological kinds fail to reduce to physical kinds, then inductions about biological kinds will hold in virtue of the essential properties of such kinds. In either case, there will be sound inductive inference about biological kinds.

4 Alternatively, the objection might be put as follows: there is a profound difference between kinds in the life sciences and kinds in physics. The present account applies to kinds in the physical sciences. Yet any account of induction must also apply to kinds in the life sciences.

Chapter 6

Methodological Pluralism, Normative Naturalism and the Realist Aim of Science

6.1 Introduction

There are two chief tasks which confront the philosophy of scientific method. The first task is to specify the methodology which serves as the objective ground for scientific theory appraisal and acceptance. The second task is to explain how application of this methodology leads to advance toward the aim(s) of science. In other words, the goal of the theory of method is to provide an integrated explanation of both rational scientific theory choice and scientific progress.[1]

Theorists of scientific method may be broadly divided into two main camps: monists and pluralists.[2] Traditional methodologists tend to fall into the monist camp. They see science as characterized by a single, universally applicable method, invariant throughout the history of science and the various fields of scientific study. The two leading versions of monism are inductivism, which takes scientific theories to be grounded in inductive inference from observed data, and Popperian falsificationism, which treats the method of science as the ruthless attempt to refute conjectural hypotheses which scientists propose to explain observed phenomena.

---

1 This view of the task of the theory of method accords, for example, with the two ingredients of a 'rational model of scientific change' identified by Newton-Smith, viz., specification of the goal of science and of principles of theory comparison (Newton-Smith, 1981, 4). The demand for an integrated response to both tasks is well-exemplified by Lakatos's plea for a 'whiff of inductivism' in Popper's treatment of the relation between corroboration and verisimilitude (Lakatos, 1974, 256).

2 The distinction between monist and pluralist theories of method is somewhat crude, since there are also mixed positions. John Worrall, for instance, holds that there is an invariant core of methodological principles, which remains fixed throughout change of lower level principles (Worrall, 1988). The issue of methodological variance masks further complexity, as well. For, in principle, one might argue that at any one time science is governed by a single method, though this method may undergo historical variation. Conversely, one might argue that there is a plurality of methods which are historically invariant. Hence, a full taxonomy of methodological views would include variationist and invariationist versions of both pluralism and monism, in addition to mixed positions.

By contrast, recent methodological pluralists argue, against the idea of a fixed method, in favour of a plurality of methodological rules governing theory evaluation.[3] Such methodological rules may vary from time to time, as well as field to field, within science. New rules may be introduced and old ones discarded. Rules may be modified, as they undergo refinement in the course of scientific practice. They may be applied in different ways in different fields of science, and different scientists may interpret the same rules in different ways. Moreover, as there is always a plurality of rules, different scientists may choose to emphasize different rules in the evaluation of alternative theories. On the resulting pluralist conception of methodology, science is not characterized by a single invariant method, but by a set of evaluative rules to which scientists appeal in the context of theory appraisal.[4]

As for the aim of science, a number of alternative approaches may be distinguished here as well. According to scientific realism, the aim of science is to arrive at true, explanatory theories of both observable and unobservable aspects of the world, and the best explanation of the success of science is that considerable headway has been made toward that aim. For the empiricist, by contrast, the aim of science is restricted to producing predictively accurate theories which are empirically well-supported by the observed phenomena. Conventionalist philosophers of science, who regard theories as classificatory schemes which impose order on experience, take the main aim of science to be to produce an economical ordering of experience. Philosophers

3 The best-known pluralists are Feyerabend (1975), Kuhn (1977, Ch. 13) and Laudan (1984). Elements of a pluralist methodology may be found in the work of such authors as Chalmers (1982), Ellis (1990), Lacey (1997), Lycan (1988, Ch. 7), McMullin (1987), Newton-Smith (1981), Quine and Ullian (1970) and Thagard (1978). I defend a pluralist stance in the later chapters of my (1997).

4 *Terminological note*: some comment is necessary regarding my use of the term 'methodological rule' and related expressions. A variety of terms (e.g., 'criteria', 'norm', 'principle', 'rule', 'standard', 'value') is found in the methodological literature. While there are slight differences of meaning and usage, there is no substantive difference between such terms of relevance to the issues dealt with in this chapter. All such terms denote methodologically relevant factors to which appeal is made in theory appraisal and justification of theory choice. The terms might therefore be used interchangeably. However, to reduce scope for confusion I will tend instead to speak either of criteria or of rules, restricting use of related terms to contexts in which another term seems especially apt. I will understand the relation between criteria and rules to be roughly as follows: a criterion is a methodologically desirable feature of a theory (e.g., accuracy, coherence, simplicity); rules are prescriptions typically (but not necessarily) stated in linguistic form (e.g., 'avoid *ad hoc* hypotheses', 'employ double blind tests'). In general, criteria (e.g., simplicity) may be stated in an analogous form as rules (e.g., 'prefer simple hypotheses'). It is also worth noting that for present purposes no decision need be made as to whether rules or criteria are best construed as necessary and/or sufficient conditions of theory acceptance, or merely as factors of relevance to theory appraisal. Hence, I ignore as irrelevant in the present context the otherwise important distinction between rules which dictate theory choice and values which merely guide such choice (cf. Kuhn, 1977, 331).

of a pragmatist bent emphasize prediction and control of the environment, in the service of successful achievement of practical goals.[5]

In this chapter, I will focus on the relationship between methodological pluralism and scientific realism. In particular, I will consider the question of whether sustained application of a plurality of methodological rules conduces to realization of the scientific realist aim of truth. This question, which raises issues of both an epistemological and a metaphysical nature, is a special instance of the more general demand for an integrated account of rational theory choice and scientific progress. It is, in my view, *the* most urgent question facing the scientific realist who seeks to derive insights about scientific methodology from the pluralist approach found in the work of T.S. Kuhn and P.K. Feyerabend.

The chapter is organized as follows. In Section 6.2, I discuss the threat of relativism which is raised by methodological pluralism. In Section 6.3, I show that Laudan's normative naturalist metamethodology removes the threat of relativism. In Section 6.4, I propose that normative naturalism be incorporated within the framework of scientific realism. Section 6.5 presents objections due to Laudan against the realist aim of truth, which threaten the incorporation of normative naturalism within a realist framework. In Sections 6.6 and 6.7, I defend the realist aim of truth against these objections. Finally, I argue in Section 6.8 that use of a plurality of methodological rules promotes the realist aim of science.

6.2 Pluralism and Relativism

The main impetus for a pluralist conception of method derives from the historical philosophy of science notably championed by Kuhn and Feyerabend. By contrast with earlier monist orthodoxy, advocates of the historical approach argued that science should be conceived as a developmental process, which takes place in a variety of historical circumstances using a variety of methods, rather than the implementation of an invariant, universal method. Kuhn, who initially argued that standards of theory appraisal vary with scientific paradigm, later came to argue that science is governed by a set of cognitive values (e.g., accuracy, breadth, simplicity, coherence, fertility) which guide theory choice. Feyerabend, for his part, argued not only that all methodological rules are routinely violated in the course of scientific practice, but that there are often good grounds for the violation of such rules.

Some writers suppose that the historical approach of Kuhn and Feyerabend entails wholesale rejection of scientific method. However, I prefer to draw a more positive moral. What is to be rejected, if one adopts the historical approach, is not method as such, but a monistic *theory* of method. Ample scope remains to develop a more adequate account of method within the framework of the historical approach. In particular, what emerges from the historical approach is a pluralist conception of

5 The relationship between aims and methods is not straightforward. There is scope for a variety of different accounts of such relationships. For example, in contrast with other theories of method, the conventionalist elevates the aim of overall theoretical simplicity into the paramount methodological principle of science. On the other hand, realists and empiricists may agree on the nature of method but disagree on the aims served by the method.

method, on which the principles of method are not unique and invariant, but multiple and subject to variation in the history of science.

I have elsewhere attempted to sketch the main outlines of a pluralist theory of method (1997, Ch. 7). For present purposes, it suffices to characterize the pluralist account by means of the following five theses, which represent central themes of the historical school:

1. *Multiple rules*: scientists utilize a variety of methodological rules in the evaluation of theories and in rational choice between alternative theories.[6]
2. *Methodological variation*: the methodological rules utilized by scientists undergo change and revision in the advance of science.[7]
3. *Conflict of rules*: there may be conflict between different methodological rules in application to particular theories.[8]
4. *Defeasibility*: the methodological rules, taken individually rather than as a whole, are defeasible.[9]
5. *Non-algorithmic rationality*: rational choice between theories is not governed by an algorithmic decision procedure which selects a unique theory from among a pool of competing theories.[10]

These five theses constitute the basic elements of a pluralist conception of methodology, according to which scientific theory appraisal is governed by an evolving set of methodological rules. Because the rules may conflict in practice, and are individually defeasible, appeal to the system of rules need not uniquely determine the outcome of theory choice. Accordingly, scientists who place differential weight on various rules may come thereby to decide in favour of opposing theories.

6 See, e.g., Ellis (1990, 244-59), Kuhn (1977, 321-2), Lacey (1997, 31-3), Laudan (1984, 33ff; 1996, 18), Lycan (1988, 129-30), McMullin (1987, 53-4) and Newton-Smith (1981, 226-32).

7 See, e.g., Feyerabend (1978, 33-9, 98), Kuhn (1970a, 103-10, 148; 1977, 335-6), Chalmers (1990, 20) and Laudan (1984, 39-40, 57-9, 81-2; 1996, 17).

8 E.g., simplicity may favour one theory, coherence or breadth another (cf. Kuhn, 1977, 323-4; Thagard, 1978, 92). For qualification of the view that there may be conflict between rules, see Laudan (1996, 93-4).

9 That methodological rules are defeasible is, of course, the main thrust of Feyerabend's opening argument in his (1975). However, the defeasibility of all rules, taken singly, does not entail that all such rules may be concurrently violated. Hence, while any particular rule may be violated in appropriate circumstances, it is rationally unacceptable to transgress the entire system of methodological rules. While perhaps not entirely explicit in Kuhn, the inviolability in general of the set of rules is in the spirit of Kuhn (1977, Ch. 13). For related discussion, see Laudan (1996, 101-5.).

10 Explicit rejection of an algorithm of theory choice occurs in Kuhn (1970a, 200; 1977, 326), and Laudan (1984, 5-6; 1996, 17-9). Chalmers tacitly denies an algorithm of theory choice in his discussion of Feyerabend's critique of universal methodological rules (1982, 135). Brown develops a non-algorithmic conception of rationality in his (1988). Explicit formulations aside, however, rejection of an algorithm of theory choice is virtually the defining thesis of the historical school.

It is precisely the scope that methodological pluralism affords for rationally grounded disagreement between scientists that makes it controversial. For it brings it into tension with methodological monist accounts which restrict rational divergence of opinion to that allowed by compliance with a single method. The opposition between monist and pluralist accounts of method is at the root of much recent concern with epistemological relativism in the philosophy of science. For, on the one hand, it is widely held that a monistic theory of method avoids relativism by grounding theory choice in a shared, invariant method.[11] On the other hand, it is also widely assumed that pluralism entails relativism, since the existence of a plurality of methods would provide scientists with rational justification for the acceptance of opposing theories on the basis of alternative sets of rules.

However, it is a mistake to suppose that rational disagreement due to variation of methodological rules necessarily leads to relativism. For that would be to suppose that mere difference in the rules employed by scientists entails relativism. And that in turn would be to suppose that mere compliance with operative rules suffices for rational justification. Yet the latter assumption is surely mistaken. It overlooks the crucial distinction between rules which provide rational justification and those which do not. Not all methodological rules that may be proposed or employed are capable of providing rational justification. Some provide no justification at all. Given the distinction between rules which provide justification and those which fail to do so, relativism is not entailed by pluralism, since mere satisfaction of a methodological rule does not suffice for rational justification.

Yet, while a plurality of methodological rules may not entail relativism, the challenge of relativism now arises in a novel form. For the distinction between rules which provide justification and those which do not is a distinction that is itself in need of defence. After all, how can one rule be shown to provide greater rational justification than another? The relativist challenge, therefore, is to show how one methodological rule may be epistemically superior to another.

6.3 Naturalism and Relativism

The question of how to assess the epistemic merits of a methodological rule is a metamethodological question about the nature of epistemic normativity. One of the most promising approaches to this issue is a form of epistemic naturalism which grounds normativity in the facts of inquiry.[12] This approach involves two key

11 As such, however, monism need not be immune to the challenge of relativism, since the question may always be raised of the justification of the monist's purportedly invariant method, as against another possible method. For relevant discussion, see Laudan (1996, Ch. 9) and Worrall (1988; 1989), as well as my (1997, Ch. 10).

12 Epistemic naturalism is not, of course, the only approach to epistemic normativity. Among the main alternatives to naturalism in metamethodology, it is worth noting the conventionalism of Popper (1959), the intuitionism of Lakatos (1978) and early Laudan (1977), and reflective equilibrium models which trace back to Goodman (1955). For further analysis of the range of metamethodological approaches, see Nola (1987; 1999) and Nola and Sankey (2000; 2007).

elements. On the one hand, it treats methodological rules as empirical hypotheses about how to pursue inquiry, which may be evaluated in light of empirical evidence. On the other hand, such rules are conceived as instruments or tools of inquiry, the epistemic function of which is to advance cognitive ends. The two elements are combined by grounding evaluation of methodological rules in empirical evidence about performance of epistemic function.

As a special case of this naturalist approach to epistemic normativity, I turn to the *normative naturalist* metamethodology of Larry Laudan.[13] Laudan is critical of the scientific realist view defended here that the aim of science is advance on truth. In the sections to follow I will explore the possibility of incorporating Laudan's normative naturalism within a scientific realist framework. However, in this section my concern is with the normative naturalist account of epistemic normativity as a response to relativism.

As a naturalist, Laudan treats metamethodology as an empirical discipline continuous with natural science. In order to ground methodology empirically, it must be possible to treat methodological rules as normative claims about the conduct of inquiry which are capable of empirical evaluation. Accordingly, Laudan proposes that methodological rules be construed in instrumental fashion as recommendations of means of realizing desired cognitive ends. This enables such rules to be formulated as conditional claims with the following hypothetical imperative form:

> If one wishes to attain aim A, then one ought to employ method M.

As an example of how a methodological rule may be cast in hypothetical form, Laudan offers the following formulation of Popper's rule against *ad hoc* hypotheses:

> [I]f one wants to develop theories which are very risky, then one ought to avoid *ad hoc* hypotheses. (Laudan, 1996, 133)

Such an analysis permits the recommendation of a methodological rule to be based on historical evidence. For it reveals how such rules may be supported by claims of statistical covariance between past use of method and achievement of results. Where use of a method has historically proven to be a reliable means of achieving a given end, the method may be recommended on the basis of past performance as means to that end. In this manner, empirical evidence from the history of science may serve as the normative ground of a methodological rule.[14]

---

13 See Laudan (1996, Ch. 7). While Laudan's normative naturalism is well-suited for the present purpose of defeating the relativist, it is but one instance of a widespread form of epistemic naturalism. Similar views of both the nature and evaluation of methodological rules may be found in Rescher (1977) and Stich (1990). The idea that methodological rules are tools of inquiry has deep pragmatist roots, which may be traced back, for example, to Dewey's comparison of methods of inquiry with methods of farming (Dewey, 1986, 107-8). Closely related views occur as well in Giere (1989) and Kornblith (1993b).

14 The role here attributed to cognitive ends by Laudan raises the spectre of a relativism due to variation of ends (cf. Psillos, 1997, 707). However, Laudan's hypothetical imperative account of rules needs to be understood in the context of his remarks on rational adjudication of cognitive goals in his (1984, 50ff). Laudan there adumbrates a number of means of evaluating

The normative naturalist analysis of the justificatory basis of methodological rules enables the distinction to be sustained between rules which provide epistemic support and ones that do not. For if use of one rule reliably conduces to a given aim and use of another fails to, then it provides greater epistemic support than the other. But if one rule may have greater epistemic merit than another, the challenge of relativism may be met. For where there may be variation in the epistemic credentials of rules, rational justification does not reduce to mere compliance with operative methodological rules. Hence, one theory may enjoy a higher degree of support than another, despite a plurality of methodological rules.

6.4 Scientific Realism and Normative Naturalism

Laudan is a well-known critic of the realist view that truth is the aim of science. Accordingly, Laudan develops normative naturalism within the context of an axiology that allows a multiplicity of scientific aims, rather than being limited to the realist aim of truth. However, in contrast with Laudan, I seek to combine methodological pluralism with scientific realism precisely by incorporating normative naturalism into a realist framework. In so doing, I wish to preserve the normative naturalist response to epistemic relativism while providing an integrated account of both the methodology of science and its progress.

The core of the normative naturalist analysis of methodological rules is that rules may be construed as hypothetical imperatives linking epistemic means and ends. This enables such rules to be treated instrumentally as cognitive tools, which may be utilized to advance the aims of science. Such an instrumental analysis of methodological rules leaves the nature of the epistemic or cognitive aims unspecified. As a critic of realism, Laudan rejects the realist aim of truth, for reasons to be considered in the next section. However, Laudan does not offer any one, unique alternative to truth as the correct analysis of the constitutive aims of science. Rather, he argues that scientists' cognitive aims vary historically as part of the continual process of adjustment and correction of theories, methods and aims which characterizes scientific inquiry.[15]

Because the instrumental analysis of rules is neutral with respect to the nature and number of aims that scientists may pursue, I hold it to be possible to set the analysis within a realist framework. In particular, if we treat truth as the paramount aim of science, we may then suppose that the cognitive aim that is to be fulfilled by a proposed rule is advance on the truth about the world.[16] On such a realist construal

---

cognitive aims, e.g., by showing an aim to be utopian, or in conflict with practice. It should be allowed, therefore, that Laudan seeks to avoid relativism due to variation of cognitive aims. Whether he succeeds is another matter.

15 As examples of cognitive aims that have been pursued by scientists, Laudan mentions infallible knowledge, high probability, simplicity, elegance, as well as Newton's attempt to reveal divine agency at work within the physical world (cf. Laudan, 1984, 51ff; 1996, 129).

16 To say that science aims for truth is not to be distinguished from saying that it aims for truth about the world. Nor would I distinguish it from saying that the aim of science is knowledge, since knowledge implies truth. Nor either would I demur if a realist were to argue

of normative naturalism, a methodological rule conveys epistemic warrant to the extent that fulfillment of the rule conduces to the aim of truth. As such, normative naturalism emerges as a species of reliabilist epistemology once it is placed within the context of scientific realism. For it is reliability in leading to the truth which is then the basis of the epistemic warrant of methodological rules.[17]

Where the realist sees truth as the aim of science, Laudan allows that a multiplicity of aims may be pursued by scientists. However, in speaking of truth as the aim of science, the realist need not deny that scientists pursue multiple aims. Instead, the realist need only conceive truth as the paramount aim that constitutes the ultimate goal of science. The various other cognitive aims which may be pursued by scientists may be understood as subordinate aims which subserve the overriding realist aim of truth. This permits the realist to preserve an additional aspect of Laudan's analysis of the epistemic warrant of methodological rules. Where Laudan holds that the warrant of a rule consists in reliable promotion of cognitive ends, the realist need not insist that the specified aim of the rule be truth. Rather, provided that the specified aim subserves the overriding goal of truth, a rule which immediately conduces to a lower level aim may still convey epistemic warrant.[18]

On the assumption that employment of methodological rules conduces to truth, or to aims that subserve truth, the present proposal offers an integrated account of both the methodology and progress of science. However, as I now turn to Laudan's objections to realism, we are about to see that this assumption is in need of defence.

6.5 Laudan and the Aim of Truth

Laudan has argued against scientific realism on a number of occasions. Perhaps most notable is his attack on convergent epistemological realism, in which he attempts to sever the explanatory connections drawn by realists between reference, truth and the success of science (1984, Ch. 5). Here, however, I focus on two specific objections

---

that the aim of science is explanation, as Ellis (1985) does, since seeking true explanations is part of seeking the truth. (However, I would demur at Ellis's suggestion that we renounce the correspondence theory of truth in favour of a pragmatist concept thereof.)

17 More specifically, combining the instrumental analysis of rules with the aim of truth yields a form of *method*, rather than *process*, reliabilism (cf. Goldman, 1986, 93-5). However, I do not wish to endorse a pure reliabilism on which warrant is strictly identified with truth conduciveness. Such an account is subject to counterexamples, such as Lehrer's case of Mr. Truetemp, who reliably forms true beliefs about the temperature due to a device implanted in his brain, but is ignorant of both the reliability of his beliefs and of their cause (Lehrer, 1990, 163). My view is roughly that reliability is a crucial part of the warrant of methodological rules, but that *use* of rules must meet additional constraints, such as being deliberately employed by a scientist on the basis of awareness of such rules.

18 As an example of a methodological rule which immediately advances a lower-order aim, and indirectly advances the aim of truth, consider Popper's rule against *ad hoc* hypotheses. Avoidance of *ad hoc* hypotheses serves to increase the falsifiability of theories, which thereby subserves the aim of truth, since the ruthless testing of falsifiable theories is held by Popper to conduce, fallibly, to truth, or at any rate to greater verisimilitude.

raised by Laudan against the realist aim of truth. These objections pose a serious threat to my proposal to set the normative naturalist account of epistemic warrant within the context of a realist account of the aim of science.

Laudan's objections turn crucially on what he takes to be the transcendent nature of truth. He assumes that we can tell neither that a theory is true nor that progress toward truth has occurred. Given this initial assumption, Laudan develops two separate arguments that truth cannot serve as a suitable aim for science. He argues, first, that it is not rational to pursue a goal which cannot recognizably be attained or even approached. Second, he rejects transcendent aims such as truth as unsuited to a naturalistic treatment of the methodology of science. Before presenting these two objections, I will examine Laudan's view of the transcendence of truth.

For Laudan, a transcendent aim or property is one to which we have no epistemic access. He describes truth as a 'transcendental property', and contrasts it with an 'immanent' goal such as 'problem-solving effectiveness', which '(unlike truth) is not intrinsically transcendent and hence closed to epistemic access' (1996, 78). The distinction between immanent and transcendent states corresponds more or less to that between what can be empirically shown to be the case and what cannot. Laudan's grounds for taking truth as transcendent appear to be twofold. On the one hand, he contrasts transcendent aims with the 'detectable or observable properties' (1996, 261, fn. 19) that provide evidence of methodological means/ends relationships, implying thereby that a transcendent state is one that cannot be directly observed to obtain. On the other hand, he claims that 'knowledge of a theory's truth is radically transcendent', since 'the most we can hope to "know" about [a theory ...] is that [it is] false' and 'we are never in a position to be reasonably confident that a theory is true' (1996, 194-5).[19] The epistemically transcendent therefore emerges as that which transcends the empirical either by being unobservable or by being based on an ampliative inference that extends beyond the observed data. Accordingly, that is what I shall mean when I speak in what follows of the transcendence of truth or theoretical truth.

Laudan accords truthlikeness a status similar to truth. Since the truth of a theory transcends our capacity for knowledge, we can be in no position to judge how closely an actual theory approximates the truth (Laudan, 1996, 78). The problem is aggravated by lack of a clear conception of approximate truth. On the Popperian account of verisimilitude, for example, a theory may have high verisimilitude and yet display little or no empirical success (Laudan, 1984, 118). More generally, Laudan

19 Laudan credits the point that we cannot know a theory to be true to Hume and Popper (1996, 194). However, he also notes (personal communication) that his point is intended to be stronger than simply saying that theories cannot be shown to be true. He refers to the latter as 'Humean underdetermination' (1996, 31). By contrast, his point about the transcendence of truth appears to be a strong version of what he describes as 'ampliative underdetermination' (1996, 43ff). For while Laudan denies that ampliative rules of inference underdetermine rational theory choice, his claim that theories cannot be reasonably held true seems to imply that such rules underdetermine rational belief *in the truth of theory*. The grounds for this thesis would appear to be either a version of the 'pessimistic meta-induction' (cf. 1977, 126) or his related critique of the explanatory connections drawn by realists between scientific success and truth (1984, Ch. 5).

claims that there is no known means to measure or estimate how close a theory is to the truth. Consequently, truthlikeness transcends our capacity to know it every bit as much as does truth.

Given the transcendence of truth and truthlikeness, Laudan objects to the role accorded such notions within realist accounts of scientific progress. He develops his first objection in the context of a discussion of the rational evaluation of cognitive goals in his (1984, 50-5). According to Laudan, a crucial consideration in evaluating a goal is whether it may be realized. He takes it as a requirement of rationality that there be grounds to suppose it possible to achieve the goals one pursues (1984, 51). Goals which are unable to be achieved may be rejected as 'utopian'. Laudan distinguishes three ways in which goals may be utopian: goals that can be shown to be unrealizable are 'demonstrably utopian'; ones that are overly vague or imprecise are 'semantically utopian'; and goals which cannot be shown to obtain are 'epistemically utopian'. Laudan's objection to truth as a cognitive goal is that it is epistemically utopian.

As a prime instance of an epistemically utopian goal, Laudan takes the 'goal of building up a body of true theories' (1984, 53). He allows that such a goal may not be demonstrably utopian, and that the concept of truth admits of clear analysis. However, he asks us to consider the case in which one 'has no idea whatever how to determine whether any theory actually has the property of being true' (1984, 51). (Of course, as we have just seen, Laudan takes this to be our actual epistemic situation, given the transcendence of truth.) In such a case, where value is placed on an unrecognizable property, Laudan says that 'such a value could evidently not be operationalized' (1984, 53), meaning by the latter that no procedure is known which would lead to its attainment (cf. 1984, 51). He then concludes that:

> if we cannot ascertain when a proposed goal state has been achieved and when it has not, then we cannot possibly embark on a rationally grounded set of actions to achieve or promote that goal. In the absence of a criterion for detecting when a goal has been realized, or is coming closer to realization, the goal cannot be rationally propounded even if the goal itself is both clearly defined and otherwise highly desirable. (1984, 53)

Given that Laudan takes truth and truthlikeness to be transcendent, I suggest he is to be understood as proposing the following argument against the realist aim of truth: (a) it is not rational to pursue an aim which may neither be recognized to obtain nor to be close to obtaining; (b) the goal of true theories may neither be recognized to obtain nor to be close to obtaining; therefore (c) it is not rational to pursue the goal of true theories.[20]

While Laudan's first objection concerns rational pursuit of truth, his second objection derives from his naturalistic view of method. In particular, Laudan argues that transcendent goals such as truth are shown to be illegitimate by the normative

---

20 It might be objected that Laudan states the argument in conditional form, e.g. 'if we cannot ascertain when a proposed goal state has been achieved'. Hence, it is not to be interpreted as an argument against realism, but merely as an example of a possible epistemically utopian aim. However, since, as we have seen, Laudan holds truth to be transcendent, he is committed to dismissing it as an epistemically utopian aim, which cannot be rationally pursued.

naturalist analysis of methodological rules. As we saw in Section 6.3, the normative naturalist construes methodological rules in instrumental fashion as hypothetical imperatives which relate cognitive means and ends. Such an analysis enables methodological rules to be evaluated empirically with regard to their effectiveness in promoting specified aims. According to Laudan, the instrumental conception of method places a premium on the realizability of aims. Aims which cannot be achieved (i.e., utopian aims) are unsustainable, given the goal-directed nature of methodology.

More specifically, Laudan claims that the instrumental conception of method leads to rigorous constraints on the legitimate aims of science:

> any proposed aims for science [must] be such that we have good reasons to believe them to be realizable; for absent that realizability there will be no means to their realization and thus no prescriptive epistemology that they can sustain ... (1996, 157-8)

Such constraints have direct bearing on the realist aim of truth:

> one of the corollaries of the instrumental analysis is that those ends that lack appropriate means for their realization become highly suspect. Traditional epistemologists who ... hanker after true or highly probable theories as the aim of science find themselves more than a little hard pressed to identify methods that conduce to those ends. Accordingly, normative naturalism suggests that unabashedly realist aims for scientific inquiry are less than optimal. (1996, 179)

Thus, the demand of realizability entails the rejection of realist aims as unacceptable for science. The reason, as with the previous objection, turns on the transcendent nature of truth:

> if one has adopted a transcendental aim, or one which otherwise has the character that one can never tell when the aim has been realized and when it has not, then we would no longer be able to say that [a] methodological rule asserts connections between detectable or observable properties. I believe that such aims are entirely inappropriate for science, since there can never be evidence that such aims are being realized, and thus we can never be warrantedly in a position to certify that science is making progress with respect to them. (1996, 261, fn. 19)

In short, because methodological rules derive their epistemic support from underlying empirical means/end connections, there may be no evidence capable of showing that a rule promotes a transcendent aim, since no empirical evidence may show that a transcendent aim has been reached or is close to being reached.

Based on the lack of possible evidence for advance on truth, Laudan concludes that the realist aim of truth fails to be a legitimate goal for science. While it is not entirely clear how the various strands of Laudan's thoughts on this topic fit together, I propose the following reconstruction of his argument: (a) the methods of science are instruments for the realization of the aims of science; (b) given this, a legitimate aim of science must be such that it may be realized *and* there may be evidence of its realization; (c) because truth is transcendent there may be no evidence that the end of truth is realized; hence (d) truth is not a legitimate aim of science.

In sum, Laudan rejects the realist aim of truth on the grounds that it is neither rational to pursue the truth nor is the truth a legitimate aim of science. Both of these objections turn on the basic assumption that truth is transcendent. Let us now see if these objections may be met.

6.6 Is Truth Transcendent?

The two objections canvassed in the preceding section stem from the common premise that theoretical truth is transcendent. In this section I will challenge this premise by arguing that it is possible to have theoretical knowledge. In the next section, I will address the negative consequences which Laudan derives from the premise about the rationality and legitimacy of pursuit of truth.

As we have seen, Laudan regards theoretical truth as transcendent in the sense that such truth transcends our capacity to know it. However, it is by no means evident that theoretical truth is unknowable, as Laudan claims it to be. That this is so may be readily shown on the basis of the standard analysis of knowledge as justified true belief. On such an analysis, a knowing subject S knows a theoretical proposition P if and only if three conditions are fulfilled:

1. S believes that P is true,
2. S's belief that P is true is rationally justified,
3. P is true.

Given such an analysis of knowledge, there is no apparent reason in principle why a theoretical proposition may not be known to be true. For in order to know that P is true, it suffices that there be good grounds for the belief that P and that P in fact be true.

More specifically, let us suppose that a scientist believes a theoretical proposition P (e.g., 'Electrons have negative charge') to be true. On the assumption that it is possible for a theoretical proposition to correctly report an actually existing state of affairs (e.g., that electrons in fact have negative charge), then it is possible for P to be true. Provided, moreover, that P satisfies appropriate methodological standards, there may be good rational grounds for the belief that P is true. Given both these assumptions, and the standard analysis of knowledge, it follows that P may be known to be true, for one may rationally believe P and P may be true. Hence, theoretical knowledge is possible.

Against this, it might be objected that one may have a justified true belief that P and yet be unable to tell that P is true. The objection arises because P is a theoretical proposition whose truth is not directly evident. For, while P may well be true, there is no direct means of knowing that this is so. At most, one may have access to the evidence which justifies the belief that P. But there is no access to the truth of P that is independent of the evidence for P. Thus, even if P is true, and justifiably believed to be so, one may fail to be in a position to know that it is true. Given this, the fact that the conditions specified for knowledge may be fulfilled in the case of a theoretical proposition does not show that theoretical knowledge is possible.[21]

This objection rests on a confusion between conditions for the possession of knowledge and criteria for the recognition of knowledge. The justified true belief analysis of knowledge provides a set of conditions, satisfaction of which qualifies a subject as having knowledge. It does not provide criteria which enable a subject to recognize that those conditions obtain, and is thereby in possession of knowledge. Thus, it is possible for one to know that P without being able to recognize that one knows that P or that P is true. In short, one may have theoretical knowledge even in the absence of direct epistemic access to the truth of the theoretical proposition that is known.[22]

Such absence of direct access leads to a further potential objection to theoretical knowledge. For if there are no criteria which enable recognition of theoretical truth, then such truth may not be shown with certainty to obtain. One might then object that theoretical knowledge is not certain knowledge, and so not strictly knowledge at all. Such an objection is suggested by Laudan's previously quoted discussion of the 'epistemically utopian' character of truth, where he says that the value of truth cannot be 'operationalized' and that there is no 'criterion for detecting when a goal [e.g., truth] has been realized' (1984, 53). However, I am loath to attribute this objection to Laudan, since he is on record as supporting fallibilism (e.g., 1984, 51-2; 1996, 213), and indeed dismisses 'apodictic certainty' as a transcendent property on a par with truth (1996, 78).[23] In any event, it is a commonplace of the philosophy of science that scientific theories are constantly subject to revision with the advance of science, so that any adequate conception of scientific knowledge must allow that one may have knowledge without certainty.

There remains an additional basis on which to object to the possibility of theoretical knowledge. Laudan might object to the present use of the justified true

21 The present objection to the standard analysis differs from Gettier-style objections. Gettier cases show that the standard analysis fails to provide a set of jointly sufficient conditions for knowledge. By contrast, the present objection turns on lack of direct epistemic access to the truth of theoretical propositions. Incidentally, while Gettier cases show that further conditions are needed to obtain sufficient conditions for knowledge, the conditions specified by the standard analysis remain individually necessary and thereby constitute an approximately correct analysis of the concept of knowledge. Given this, it is unproblematic to treat the standard analysis as an adequate working definition of knowledge.

22 This implies the falsity of the KK-thesis, i.e., the thesis that in order to know one must know that one knows. I take the KK-thesis to be false, since one may know without being aware that one knows, or even knowing what it is to know.

23 However, it is not completely clear what Laudan takes to follow from fallibilism with respect the concept of knowledge. He writes at one point that 'the unambiguous implication of fallibilism is that there is no difference between knowledge and opinion: within a fallibilist framework, scientific belief turns out to be just a species of the genus opinion' (1996, 213). This might be taken to suggest that knowledge has no greater warrant than any other form of belief. However, since, in the context in question, certainty is the crucial factor which distinguishes opinion from knowledge, knowledge might still be justified true belief and yet belong to the genus opinion.

belief analysis of knowledge on the basis that there may be no grounds which could rationally justify a scientist in believing that a theoretical proposition is true.[24] In other words, he might deny that the grounds which provide rational support for a theoretical proposition provide support for the truth of the proposition. At first blush, this may seem an implausible objection, since, as has been noted by a number of authors, rational grounds for belief that P are *ipso facto* rational grounds for the belief that P is true.[25] For if one has grounds for the belief that P, then, by semantic ascent, one has grounds for the belief that P is true. Hence, one cannot sever rational belief from rational belief in truth in the manner that the objection requires.

There is, however, a consistent line of argument available to Laudan here. On the instrumental analysis of rules, the warrant of a methodological rule relates to the end served by the rule. Hence, since there may be no evidence that a rule conduces to theoretical truth, satisfaction of a rule may provide no warrant for belief in such truth. Rather, satisfaction of a rule provides warrant only with respect to the end served by the rule. Thus, when the aim served by a rule is that of predictive reliability, for example, satisfaction of the rule by a theory licenses belief that the theory is predictively reliable, not that it is true. Given that justification always relates to the end served by a rule, it is therefore consistent for Laudan to hold that there may be rational grounds for a theory that are not grounds for believing that the theory is true.

However, while it may be consistently denied that a warrant need be a warrant for truth, the resulting position is unsustainable for several reasons. For one thing, it leads to an implausible restriction on the epistemic states of scientists. For if there may be no warrant for belief in theoretical truth, no scientist who accepts any theory as true may do so rationally, no matter how weighty the evidence or how well-established the theory. For another thing, it rests on an unduly narrow empiricist epistemology.[26] For if there may be no warrant for belief in the truth of any proposition that transcends empirical evidence, then all inferential or indirect knowledge is precluded due to lack of rationally justified belief. Finally, denial that methodological criteria provide warrant for truth removes the rationale for scientists' use of a plurality of such criteria in the evaluation of theories. Scientists who accept a theory which satisfies multiple criteria (e.g., predictive accuracy, explanatory breadth, simplicity, coherence) may do so because they interpret such joint satisfaction of criteria as indicating the likely truth of the theory. But in the absence of such a unifying aim served by criteria, scientists are deprived of a rationale for conjoint use of multiple criteria.

I conclude that there is every reason to suppose that theoretical knowledge is possible. Neither our lack of direct or infallible epistemic access to theoretical truth,

---

24 That this is indeed Laudan's likely objection is suggested by my footnote 19 (above).

25 The point is made specifically with regard to Laudan by Psillos (1997, 712). Lycan makes the point in a more general context in response to the claim that one may have evidence for P but not evidence for the truth of P (Lycan, 1988, 137).

26 The point that Laudan's epistemology is unduly empiricist has been made by a number of authors, including most relevantly (Nola, 1999, 10). It should be noted that Laudan explicitly denies the charge (1996, 160). But his denial is difficult to reconcile with his dismissal of theoretical truth as a 'transcendent' aim.

nor the possibility of a warrant that is not a warrant for truth, entails that we are unable to have theoretical knowledge. It may not be possible to prove beyond a shadow of a doubt that a theoretical proposition is true. But that does not mean that such truth radically transcends our epistemic capacities, as Laudan suggests.

6.7 The Pursuit of Truth

In this section I will consider Laudan's two objections to truth as the aim of science. As we saw in Section 6.5, Laudan argues that truth is epistemically utopian, hence unable to serve as an object of rational pursuit. Nor is truth admissible as an aim of science, since there may be no evidence of its realization. Since both objections depend on the transcendence of theoretical truth, they are in large part undermined by the possibility of theoretical knowledge for which I argued in the previous section. However, it remains to show this in detail.

If theoretical knowledge may be acquired by methods employed by scientists, it would seem natural to suppose that acquisition of such knowledge is a legitimate goal for science. Before further scrutinizing this assumption, however, I will briefly consider the consequences of denying that theoretical knowledge is possible. One might think that if theoretical truth or knowledge were wholly unattainable, there could be no rationale for their pursuit. For it is futile to attempt the impossible.

However, as Rescher notes against Laudan, there are circumstances in which it is rational to pursue an unattainable ideal (Rescher, 1982, 227). Moral perfection may be beyond our reach, for example, but striving for such perfection may make one a better person. Similarly, truth may function in the manner of a 'regulative ideal' for science. For, while it may be impossible for science to achieve perfection, the idea of a perfectly true theory may serve to maintain the self-corrective, evolutionary character of the scientific enterprise. In addition, the pursuit of an unattainable ideal may yield indirect benefits which are themselves otherwise unattainable. For example, it is arguably the case that the ideal of a comprehensive, true theory of the world exerts pressure on science to develop systematic theories with real explanatory breadth. Indeed, such lower level values as explanatory breadth would seem to have little independent rationale in the absence of a demand for a comprehensive, true theory.

The possibility of a regulative role and indirect benefits secures for truth a legitimate place in science even if it is unattainable by scientific means. However, if, as argued in the previous section, theoretical knowledge is possible, then truth is in fact an attainable end that lies within the reach of science. This would seem to vindicate theoretical truth as a legitimate goal of rational scientific inquiry. For, on the one hand, if truth is a realizable aim of science, it is possible for an agent to rationally pursue truth as a goal. On the other hand, the attainability of truth means that it satisfies the requirement of the instrumental conception of method that only achievable aims be allowed into science.

But Laudan's principal objection is not that theoretical truth is inappropriate as an aim because it cannot be attained. His main point is that we would be unable to recognize truth even if we were to attain it. Given this, it is not rational for an agent to pursue truth, since there are no criteria which would enable one to recognize attainment of the aim or that it is close to attainment. Similarly, it is because there may be no evidence indicating that a method yields truth that truth is excluded as an admissible aim of science.

Laudan's emphasis on the absence of criteria for the recognition of truth may suggest that he endorses the requirement, rejected in the previous section, that one must be able to recognize that one satisfies the conditions of knowledge in order to possess knowledge. But, in fact, Laudan's claim is not that ability to recognize truth is a requirement of knowledge. Rather, his claim is that it must be possible for one to recognize the fulfillment of an aim in order to rationally pursue that aim. Thus, his objection to the rational pursuit of truth is not that we are unable to possess theoretical knowledge because we cannot recognize truth. It is that we are unable to recognize whether an action furthers an aim, where the aim happens to be truth. Laudan therefore takes ability to recognize achievement of an aim as a requirement for the rational pursuit of that aim, not as a requirement for knowledge.

But, while Laudan may only require recognition criteria for rational pursuit rather than knowledge, similar considerations apply in either case. For Laudan's denial that there are criteria for the recognition of truth is only plausible on the assumption that such criteria must provide an infallible indication of truth. It may readily be conceded that there are no infallible criteria of truth. But it by no means follows that there are no fallible criteria for the recognition of truth. While satisfaction of methodological criteria cannot decisively prove a theory to be true, it may provide good grounds for believing a theory to be true or close to truth. There may well be no criteria which enable a rational agent to know with certainty that they are advancing on truth or have attained it. Nevertheless, such an agent may justifiably believe that a theory which better satisfies the criteria than a rival theory is likelier to be true, or closer to truth, than the alternative theory. Given this, it is entirely possible for an agent to rationally pursue the goal of truth, since satisfaction of methodological criteria may provide a fallible indication of advance on that aim.

Similar remarks apply to Laudan's objection that truth is an inadmissible aim for science, since there may be no evidence that truth is realized by any method. As we saw in Section 6.5, the objection derives from Laudan's instrumental conception of method. What motivates the objection is the thought that if a method functions in the manner of an instrument, then it is to be assessed by how well it brings about the end for which it is proposed. If there is no evidence that it performs its function, then it may not be proposed as a means to that end. The question is whether it is fair to suppose that there may be no evidence that a method leads to truth. It is perhaps true that there may be no direct empirical evidence that use of a method leads to theoretical truth. But there may surely be indirect evidence that a method conduces to such truth. For where the lower level ends served by a method are ends which themselves may be taken to subserve the aim of truth, the success of the method in conducing to such lower level ends may be taken as evidence that the methods conduce to truth. Just as there may be no infallible criteria for the recognition of truth, there may be no infallible evidence that use of a method serves truth. But that is only to say that there is no certain knowledge in theoretical matters.

Finally, a brief remark is in order regarding the basis of the objection. The objection is based on the instrumental conception of method. This gives rise to the demand for realizability. But no independent argument is given for the instrumental conception, other than that it permits empirical evaluation of methodological rules within a naturalist framework. This is admittedly a powerful point in its favour. But, if the instrumental conception really does entail that truth is an unacceptable aim for science, this may equally well be regarded as a mark against the instrumental conception. In other words, the fact that the instrumental conception excludes truth as an allowable aim may be taken to count against the instrumental conception rather than against the aim of truth. However, since I remain unconvinced that the prospects of finding a place for truth within normative naturalism are as dim as Laudan claims, I see no need at this juncture to put the instrumental conception in question. Besides, as will be seen in the next chapter, I hold that the instrumental conception of method may in fact be put to realist use as part of a realist theory of method.

6.8 Conclusion

In this chapter I have sought to show that a normative naturalist account of epistemic warrant may be combined with a scientific realist conception of the aim of science. On the general picture which emerges, the naturalistic basis of a non-relativist methodological pluralism may be sustained within a scientific realist framework. As such, the present approach affords a unified account of the method of science and its progress. However, since methods may cohere with aims without promoting them, it remains to show that use of a plurality of methodological criteria advances the realist aim of truth.

Some philosophers deny that there is a problem relating method to truth. Internal realists define truth as maximal (or ideal) satisfaction of methodological criteria. For internalists, advance on truth is the inevitable result of the use of criteria. Truth is not something separate from method to which its use may or may not give rise. Rather, for internalists, continued application of methodological criteria produces theories which increasingly satisfy the criteria. The result is advance on truth, since truth simply is maximal satisfaction of the criteria.

As a realist, I hold that the objective world in no way depends on thought. Therefore I do not equate truth with satisfaction of methodological criteria.[27] The relation between method and truth is not an internal or conceptual relation. It is an external or synthetic relation. The sole question is whether the relation is necessary or contingent. In Chapter 5, I defended the view that the epistemic warrant of certain enumerative inductions rests on the essential properties of natural kinds of things. But while I hold that metaphysical necessity grounds the reliability of certain basic kinds of inductive inference, I do not see an analogous role for metaphysical necessity in the case of theory appraisal since the latter involves factors beyond those involved in basic induction. I take the relation between method and truth to be a contingent

27 My reason is that epistemic theories of truth such as internal realism entail the mind-dependence of reality (see Devitt and Sterelny, 1987, 195-6 and Musgrave, 1999, Ch. 10). The matter is discussed at some length in 7.6 and 8.2.

relation between epistemic means and ends, which may be known in the *a posteriori* manner suggested by Laudan's naturalist metamethodology.

However, as Laudan notes, no direct empirical evidence may show that use of a methodological rule yields theoretical truth. This raises the question why use of criteria of theory appraisal should be taken to promote the goal of truth. In the absence of direct evidence linking method to truth, the grounds for such a link may be at best abductive ones. More specifically, the realist claim that application of a plurality of methodological criteria leads to progress toward truth rests on an inference to the best explanation of scientific success. What best explains why scientific theories increasingly exhibit the epistemic virtues highlighted by methodological criteria is that such theories are increasingly close approximations to the truth.

In arguing this way, I seek to extend the argument of McMullin (1987) that we are warranted in taking a theory to be 'approximately true' if it exhibits 'a high degree of explanatory success' (1987, 59). McMullin takes the explanatory success of a theory to be determined by how well it satisfies the various methodological criteria of theory appraisal (1987, 54). Where a theory exhibits a high degree of explanatory success, as indicated by satisfaction of the criteria, there are good grounds to take the general kinds of entities postulated by the theory to really exist, as well as what the theory says about such entities to be broadly correct, though open to further development (1987, 59-60).

I wish to amplify McMullin's argument in two minor respects. First, I do not wish to say simply that the high degree of explanatory success of a theory, as measured by methodological criteria, permits us to infer abductively to the approximate truth of the theory. I wish, in addition, to say that where a theory possesses an impressive range of theoretical virtues (e.g., accuracy, breadth, simplicity), the best explanation of why the theory possesses such an impressive range of virtues is that it is approximately true. Second, I wish to extend McMullin's argument by explicitly applying it to the advance of science. For where a sequence of theories increasingly satisfies the methodological criteria, the best explanation is that the sequence of theories is advancing on truth. In both these ways, then, the reason for taking continued use of methodological criteria to yield advance on truth is that this best explains why our theories increasingly satisfy such criteria. It is in this sense that what is needed to bridge the gap between method and truth is an abductive argument about how best to explain scientific success. Echoing Lakatos on Popper, one might call this 'a plea for a whiff of abduction'.

Such a whiff of abduction may seem to beg the question against Laudan's critique of the realist's success argument (1984, Ch. 5). Rebuttal of that critique is, of course, beyond the scope of this chapter, but I will briefly indicate why no question is begged by the current proposal. In the first place, Laudan's critique does not impugn all use of the success argument, but only the ambitious attempt to forge a wholesale link between reference, truth and the success of science. Application on a case-by-case basis, restricted for example to entities postulated to fill specific causal roles, may escape Laudan's strictures on the success argument. In the second place, the current abduction does not proceed at the object-level from the widespread success of science to a general realist attitude toward theories, but is a metamethodological inference to an explanation of why a theory manifests a range of methodologically desirable features.

In sum, on the view I propose the realist aim of science is added to normative naturalism by an inference to the best explanation which augments lower level cognitive ends with the aim of truth. As I will argue in the next chapter (7.9-7.10), satisfaction of methodological rules provides a sound but fallible indication that a theory is on the road to truth, and may even be there already. The point is not that satisfaction of methodological rules constitutes truth. The point, rather, is that satisfaction of methodological rules is best explained by truth. To deny the reliability of the rules of method is to leave the success of theories arrived at by such rules entirely unexplained. It is to treat such success as a matter of sheer luck.

Chapter 7

Realism, Method and Truth

7.1 Introduction

Rational scientific inquiry is governed by the rules of scientific method. Adherence to the rules of scientific method warrants the rational acceptance of experimental results and scientific theory. Scientists who accept results or theories licensed by the rules of method do so on a rational basis. Thus, rational justification in science is closely connected with scientific method.

But while there is a close relation between method and rational justification, substantive questions remain about the relation between method and truth. For example, are scientists whom method licenses in accepting a theory or experimental result thereby licensed in accepting the theory or result as true? Does use of scientific method lead scientists to discover the truth about the world? Questions such as these are questions about the truth-conduciveness of method. While they relate directly to the epistemic status of method, they bear indirectly on the nature of rational justification. For if use of method conduces to truth, then, given the relation between method and justification, the warrant provided by method is warrant with respect to truth.

Questions about the relation between method and truth divide scientific realism from anti-realism in the philosophy of science. On the one side, scientific realists take the aim of science to be discovery of the truth about the world. Realists defend the view that employment of the methods of science promotes the aim of truth. On the other side, anti-realists in the philosophy of science deny the connection that realists see between method and truth. Anti-realists typically agree that method underwrites the rationality of science. Some anti-realists deny that there are good grounds for taking use of method to lead to the realist aim of truth. Other anti-realists object to the realist conception of truth, and deny that method promotes truth in the sense intended by realists.

In the present context, the key question that divides scientific realism from anti-realism about science is whether employment of method advances the realist aim of truth. This is a question about whether a proposed means for the achievement of a given end is in fact a means conducive to that end. More specifically, it is the question of whether good grounds may be given for taking the methods of science to promote the realist aim of truth.

My aim in this chapter is to defend the realist response to this question by arguing that there are strong abductive grounds for taking the methods of science to be truth-conducive. Before I turn to that task, let me first address the relation between method and rational justification in somewhat greater detail.

7.2 Scientific Method and Rational Justification in Science

I assume a traditional view of the relation between scientific method and rational justification in science. On such a view, there is a close connection between scientific method and the rational acceptance of scientific theories and experimental results. In particular, compliance by a scientist with the rules of scientific method rationally justifies the scientist's acceptance of a theory or result. A scientist whose acceptance of a theory or result fails to comply with the rules of method thereby fails to accept the theory or result on a rational basis.

However, while I assume a traditional view of the relation between method and rational justification, I do not assume a traditional view of the nature of method itself. The traditional view of method is a monistic view, according to which there is a single, historically invariant method, the use of which is the characteristic feature that distinguishes science from non-science. By contrast with the traditional monistic view, I adopt a position of methodological pluralism according to which there is a set of methodological rules which scientists employ in the evaluation of alternative theories and the acceptance of results. These rules are subject to variation in the history of science, and different rules may be employed in different fields of science. Given the plurality of rules, scientists may diverge in the rules they employ, with the result that there may be rational disagreement among scientists on matters of fact and choice of theory. On such a pluralist view of science, while no single method is characteristic of science, the sciences are generally characterized by possession of a set of methodological rules which inform the factual and theoretical decisions of scientists.[1]

Much remains to be said about the relation between method and rational justification. However, for present purposes, I will assume that the relation between method and rational justification is straightforward. The purpose of this chapter is to examine the relation between method and truth. Even if we assume that compliance with the rules of method justifies acceptance of a theory or result, the question remains of whether the theory or result is to be accepted as true. There is an epistemic gap between method and truth. My aim is to bridge this gap.

7.3 The Realist Conception of Truth

It is often said that the conception of truth best-suited to realism is a correspondence conception of truth. On such a conception, truth is a property which a statement has in virtue of a relation of correspondence that holds between the statement and the way the world is. A statement is true just in case what the statement claims to be the case is in fact the case. The relation of correspondence is, therefore, a relation

1 Methodological pluralism gives rises to the spectre of epistemological relativism. I have sought to dispel this spectre elsewhere. In this chapter, I am concerned with the relation between method and truth, rather than the nature of rational justification or the variation of the rules of method. For discussion of the methodological pluralist approach specifically as it relates to the issue of epistemological relativism, see my (1997) as well as Chapter 6 (this volume).

between language and reality. For it is a relation between a statement couched in a language and an extralinguistic state of affairs that obtains in reality.

Since a statement is true just in case the state of affairs to which it corresponds obtains, the correspondence conception satisfies the equivalence condition specified by Tarski's T-scheme:

(T) 'P' is true if any only if P.

While the T-scheme is not a definition of truth, it provides a minimal condition of adequacy that must be satisfied by any account of truth. However truth is conceived, the truth-predicate must behave in accordance with the T-scheme. Rather than a definition, the T-scheme is a schema on the basis of which metalinguistic statements of truth-conditions may be formulated for sentences of an object-language.[2] For example, replacing 'P' in (T) by 'Electrons have negative charge' yields as statement of the truth-conditions of 'Electrons have negative charge' the T-sentence:

(E) 'Electrons have negative charge' is true if and only if electrons have negative charge.

Statements such as this assert the material equivalence of sentences that predicate truth and the sentences of which truth is predicated. The T-scheme thereby specifies a correlation between the truth of statements and the states of affairs that statements report. For it stipulates that, for any sentence 'P', 'P' is true just in case a given state of affairs obtains, viz., the state of affairs that P.

But to capture the thought behind the realist conception of truth, it is not enough to say that a statement is true just in case a given state of affairs obtains. That suggests that the relation that obtains between the truth of a statement and the state of affairs that it reports might be a mere accidental correlation. But it is no accident that a statement that reports a state of affairs is true if, and only if, the state of affairs it reports does in fact obtain. For it is precisely the fact that the state of affairs obtains that makes the statement true. It is *because* electrons in fact have negative charge that the statement that electrons have negative charge is true.

Yet even if we insist that statements be made true by extralinguistic states of affairs this does not suffice for a realist conception of truth. More must be said about the nature of the extralinguistic reality that makes statements true. There are any number of non-realist positions for which statements are made true by extralinguistic states of affairs. The idealist who takes the world to be ideas in the mind of God may say that statements are made true by ideas in the mind of God. The phenomenalist who identifies reality with the permanent possibility of experience may say that statements are made true by the permanent possibility of experience. But the realist

2 The point that the T-scheme is not a definition of truth is made by Tarski: 'neither the expression (T) itself (which is not a sentence, but only a schema of a sentence), nor any particular instance of the form (T) can be regarded as a definition of truth' (1943, 110). It might perhaps be thought that a deflationary conception of truth such as Horwich's minimalism does treat the T-scheme as a definition of truth. But Horwich himself notes that deflationism 'does not provide an explicit definition, but relies on a schema to characterize the notion of truth' (1994, xv).

can accept neither the idealist nor the phenomenalist scenario. For it is a defining feature of realism that the reality investigated by science is an objective reality that is neither constituted nor determined by thought or experience.

To rule out such mentalistic scenarios, the realist must insist that what makes statements true or false are states of affairs whose existence is in no way dependent on the mental. To qualify as a realist conception of truth, the correspondence theory of truth must be supplemented with the metaphysical realist assumption of a mind-independent reality. On the realist conception of truth that results, truth consists in correspondence between a linguistically formulated statement of fact and an extralinguistic state of affairs, where the state of affairs that makes a statement true is a mind-independent state of affairs. If it is true that electrons have negative charge, then this is due to the fact that, independently of anything we think about the matter, there are electrons, and they do indeed have negative charge.[3]

7.4 The Non-Epistemic Nature of Realist Truth

The realist conception of truth is a non-epistemic conception of truth, which enforces a sharp divide between truth and rational justification. One may rationally believe a proposition that is false, just as there may fail to be rational grounds to believe a proposition that is in fact true. Far from being an absurd consequence of realism, as some may think,[4] the non-epistemic character of truth crucially underlies the central epistemological claim of scientific realism, namely that there is an epistemic gap between method and truth which is best spanned by means of realist resources.

It is important to distinguish between two different senses in which the realist conception of truth is a non-epistemic conception of truth. The first sense is a metaphysical sense, which derives from the mind-independence of the states of affairs that make statements true. The second sense is a conceptual one, which is due to the lack of a conceptual relation between truth and rational justification.

In the first sense, the non-epistemic nature of realist truth derives specifically from the mind-independent status of the truth-makers. The point turns on the ontological independence of thought and reality, rather than on any epistemic aspect

3 My insistence that the realist conception of truth requires that claims about the world be made true by mind-independent states of affairs raises the question of the status of claims about mental states and artifacts. Since minds do not exist independently of minds, and artifacts are the product of intentional human action, claims about minds or artifacts would seem incapable of being true in the realist sense. Yet presumably the realist should allow that, at least in principle, such claims might be true. To adequately address this concern would require an analysis of the concept of independence of the mental on the basis of which it may be said that claims about mental states or artifacts are made true by states of affairs that obtain independently of the mental in the appropriate sense. No such analysis can be provided here. But, fortunately, the issue may be set aside for present purposes. The kinds of claims about the world that are of principal concern here are the observational and theoretical claims of the natural sciences. I take it to be highly plausible indeed to say that such claims are made true (or made false) by the way things stand in the world independently of what we humans think about the matter.

4 Cf. Ellis (1990, 187) and Putnam (1978, 127).

of the relation between thought and reality. For the truth of claims about the world is solely determined by the existence of states of affairs which obtain independently of human thought or experience. Hence, the belief that a given state of affairs obtains does not itself – i.e., *qua* belief – have any effect on the truth or falsity of that belief. The state of affairs may obtain, or fail to obtain, whether or not anyone believes that it does. This remains the case regardless of how well justified the belief may be. Thus, given the mind-independence of the truth-makers, it is entirely possible for rationally justified beliefs about the world to be false. Indeed, given such mind-independence, the *entirety* of such beliefs might be false.

The second source of the non-epistemic character of realist truth is the lack of a conceptual relation between the concept of truth and concepts of epistemic justification. On the realist conception of truth, truth is a relation of correspondence that obtains between statements and mind-independent states of affairs that obtain in the world. A statement is true just in case an appropriate state of affairs obtains. Thus, truth depends solely on the way the world is, whether or not the world is rationally believed to be that way. As such, no epistemic condition enters into the realist conception of truth.

More specifically, to be true in the realist sense a statement need not fulfill any epistemic condition, such as evidential support or the satisfaction of methodological rules. It need only reflect the way the world is. Nor is any epistemic concept built into the realist conception of truth, since formulation of the latter makes no use of concepts of rational justification or methodology. Hence, a statement may be epistemically well-justified, in the sense of satisfying relevant methodological rules, and yet fail to be true. Indeed, a statement may be *ideally* justified and not be true, since no entailment from epistemic justification to truth is licensed by the realist conception of truth.

Both of the foregoing senses in which realist truth is non-epistemic reflect important principles of realism. The first reflects the fundamental metaphysical tenet of realism that the world investigated by science is an objective reality that lies beyond the control (though not the reach) of human thought. The second stems from the realist view that the truth of a claim about the world consists in correspondence with such an objective reality, rather than in satisfaction of criteria of epistemic evaluation.

In light of the non-epistemic nature of realist truth, the basis of the epistemic gap between method and truth is now apparent. It is not just that it is an intelligible question whether a belief warranted by the rules of method is to be accepted as true. The point is deeper than that. Because truth depends on a mind-independent reality, and is not defined in terms of epistemic criteria, a theory might fully satisfy relevant criteria and still be false. Conversely, a theory or claim about the world might be true even though it fails to fully satisfy applicable rules of method. Given the non-epistemic nature of truth, there is no logical relation between method and truth. The question must inevitably remain open whether the methods employed in science really do lead to truth.

7.5 Two Anti-Realist Strategies

I will now consider two of the principal anti-realist strategies for dealing with the relation between method and truth. Since my aim is to provide a realist bridge between method and truth, I will not attempt a detailed examination of anti-realism here. Still, to understand the realist project, it is important to contrast it with alternative approaches to the problem.

The two strategies to be considered here represent opposing anti-realist tendencies. They are the Scylla and Charybdis between which the realist must steer a course. The first strategy is that of the *internal realism* proposed by Hilary Putnam and Brian Ellis. The internal realist strategy is to bridge the epistemic gap by defining truth in terms of method, which creates an analytic relation between method and truth.[5] The second strategy, found in Bas van Fraassen and Larry Laudan, is one that I refer to as *scientific scepticism.* The sceptical strategy treats the gap between method and truth as one that cannot be bridged. It denies that satisfaction of method licenses rational belief in truth. Instead of truth, scientific sceptics offer alternative epistemic aims which they take to be achievable using the methods of science.

While detailed critique of either form of anti-realism lies beyond the scope of this chapter, it is worthwhile situating the two positions with respect to realism. By contrast with realism, the internalist denies that there is a gap between method and truth, whereas the sceptic denies that we have the epistemic means to bridge the gap. I will argue that neither anti-realist strategy yields an acceptable account of scientific knowledge of an objective world. The internalist strategy loses sight of reality, while the sceptical strategy fails to provide a sustainable account of the relation between evidence and theory.

7.6 Internal Realism

Internal realism is characterized by an epistemic conception of truth. On such a conception, truth is identified with satisfaction of criteria of epistemic appraisal. According to Hilary Putnam, for example,

> 'Truth', in an internalist view, is some sort of (idealized) rational acceptability – some sort of ideal coherence of our beliefs with each other and with our experiences *as those experiences are themselves represented in our belief system* ... (1981, 49-50)

---

5 When I speak here of the 'internal realist strategy', I mean to restrict attention specifically to the internal realist epistemic conception of truth which defines truth in terms of method or rational justification. As Brian Ellis has pointed out to me, internal realism properly understood is a substantive metaphysical position which is not restricted to an epistemic conception of truth. In particular, the internal realist position is a neo-Kantian position which denies epistemic access to a realm of noumenal objects, and treats objects, reference and reality as relative to conceptual scheme. Suffice to say that it is the relation between method and truth, rather than any more substantive metaphysical views, that are of relevance for present purposes.

Similarly, for Brian Ellis, 'truth is what is right epistemically to believe' (1990, 10). It 'is what it is ultimately right for anyone to believe, given [our natural] system of [epistemic] values' (1990, 11). Thus, according to internal realists, for a claim or theory about the world to be true is for it to be ideally justified or for it to maximize epistemic value.

For the internalist, there is an analytic or conceptual relation between method and truth. Truth consists in appropriate satisfaction of epistemic norms. Accordingly, no problem arises for the internalist of an epistemic gap between method and truth. A theory which is ideally justified, or which maximizes epistemic value, just is a true theory. Nor does any problem arise relating use of the scientific method to advance on truth. If use of scientific method leads to theories which increasingly satisfy the rules of method, it follows immediately that science advances on truth. Given that truth consists in satisfaction of the rules of method, an increase in the level of satisfaction of such rules constitutes advance on truth.

The trouble with internal realism is that it is an inherently idealist doctrine.[6] The epistemic conception of truth entails the mind-dependence of the states of affairs that make our claims about the world true. For if truth is epistemic justification, the states of affairs that make claims true necessarily fail to be objective, mind-independent states of affairs. To revert to an earlier example, suppose it is true that electrons have negative charge. For the internalist, this means that electrons have negative charge just in case we are epistemically justified in believing that electrons have negative charge. But this has the consequence that electrons only have negative charge if we are justified in believing that they do. Thus, for the internalist, the way the world is is not something that is independent of what we think. Rather, the way the world is depends on our being justified in thinking that it is a certain way. Despite promising to span the epistemic gap, internalism therefore fails to provide an account of how scientific knowledge of an objective world is possible.

7.7 Scientific Scepticism

While the internalist adopts an optimistic view of the relation between method and truth, the view of the scientific sceptic is a decidedly pessimistic one. Both van Fraassen and Laudan maintain that scientists may have good grounds for the acceptance of theories, but deny that rational credence extends to the truth of the transempirical content of theories. Thus, both authors defend a selective scepticism which denies theoretical knowledge while granting credence to observation.

For van Fraassen, the purpose of the scientific enterprise is not to discover truth, but to construct theories that are empirically adequate:

> Science aims to give us theories which are empirically adequate; and acceptance of a theory involves as belief only that it is empirically adequate. (1980, 12)

6 See Devitt and Sterelny (1987, 196) and Musgrave (1999, Ch. 10). For further discussion, see 8.2.

A theory is empirically adequate, according to van Fraassen, 'exactly if what it says about the observable things and events in the world is true – exactly if it "saves the phenomena"' (1980, 12). Van Fraassen does not deny that theories make truth-valued assertions about unobservable items. What he denies is that empirical evidence may provide support for the truth of such claims about unobservables.

For his part, Laudan holds that scientific theories may be epistemically warranted, but denies that such warrant extends to their truth. In Laudan's view, 'knowledge of a theory's truth is radically transcendent' (1996, 195). Laudan contrasts the transcendent property of truth, which he takes to be 'closed to epistemic access', with other properties which he considers to be 'immanent', such as well-testedness, predictive novelty and problem-solving effectiveness (1996, 78). The principal basis for his rejection of a warranted presumption of theoretical truth lies in his historical critique of the convergent realist claim that there is a correlation between the success of theories and their reference and approximate truth which is best explained by realist means. For Laudan, the fact that there is no way to bridge the epistemic gap between method and theoretical truth is simply a hard fact of the history of science (see Laudan 1984, Ch. 5).

The trouble with the sceptical denial of an epistemic connection between method and truth resides in the attempt to combine metaphysical realism with the possibility of a limited epistemic warrant for theories. The scientific sceptic allows that there may be epistemic grounds that warrant acceptance of a theory, but denies that such warrant extends to the truth of the non-observational content of theory. But the sceptic does not deny that scientific theories are capable of being true. Indeed, neither van Fraassen nor Laudan provide grounds for denying that there are facts about the world which make our theoretical claims about the world true or false.

But it is not possible *both* to allow that theories are made true or false by the way the world is *and* to deny that evidential support extends to the theoretical content of theories. If empirical facts about the world are capable of providing evidential support for theories, then such evidential support cannot be restricted to the non-theoretical content of theories. The reason has to do with the nature of the relationship between the empirical facts which provide support and the theories for which such facts provide support.

Scientific theories make claims about both observable and unobservable states of affairs. Among the claims which theories make about observable states of affairs are predictions of observable phenomena that are made on the basis of hypotheses about unobservable portions of reality. In the case of evidence based on the confirmation of such predictions, the predicted phenomena are events that, according to the theory, are brought about by unobservable causal processes. Because such observable events are supposed to be produced by unobservable causal processes, the evidence derived from such observable events has direct relevance to the theoretical hypotheses upon which the predictions of such phenomena are based. Indeed, given that hypotheses about unobservable processes may be the sole basis for prediction of the observable phenomena, the non-observational content of the theory is directly implicated in the evidential relation between observed fact and warranted theory.

In view of the failure of scientific scepticism to adequately account for the relation between evidence and theory, and the idealism inherent in internal realism, I conclude that neither position provides an acceptable account of scientific knowledge of an objective reality. I will now present the outlines of the scientific realist theory of the relation between method and truth that I propose.

7.8 A Realist Theory of Method

The realist theory of method that I propose consists of three key components. The first is the epistemic naturalist position that normative epistemological questions about rational justification are empirical questions about the best means of conducting inquiry into the objective natural world. The second is the position of methodological instrumentalism, according to which the rules of scientific method are 'cognitive tools' or 'instruments of inquiry', which serve as means for the realization of epistemic ends. The third is the position of abductive realism, which holds that the best explanation of the cognitive and pragmatic success of scientific theory and practice is that the rules of scientific method are genuinely truth-conducive tools, which serve as reliable means for obtaining truth.

These three elements of a realist theory of scientific method form part of a generally naturalistic, non-anthropocentric picture of the world, and of our epistemic relationship to it. We find ourselves embedded in a natural world which we did not create, and over whose fundamental character and structure we have no control. In order to survive, we must form beliefs about the world, and causally interact with it by means of action that is guided by such beliefs. Given the independence of reality from thought, the beliefs that we form about the world do not necessarily correspond to the way that the world in fact is. In such a world, we do not know in advance of inquiry how to proceed to ensure survival. Nor can we know by *a priori* means how best to pursue inquiry into the nature of reality. Thus, the question of how to learn about the world is a question about the contingent nature of our epistemic capacities and the relation of such capacities to the world. Such a question is an empirical question that can only be answered on the basis of empirical investigation into the nature of inquiry.

More specifically, on the instrumentalist conception of method that I favour, the rules of method are to be understood as means for the achievement of epistemic ends. In this I follow Larry Laudan, whose hypothetical imperative analysis of the rules of method was examined in 6.3. The instrumentalist construal of method reveals how the rules of method may be subject to empirical evaluation, since it is an empirical question whether use of a method reliably conduces to realization of a given aim. This illustrates the ability of epistemic naturalism to account for the normative force of rules of scientific method. Because the rules of method may be treated as empirically evaluable means to epistemic ends, the epistemic warrant of such rules may be grounded in empirical facts about the nature of inquiry. As such, the normativity of the rules of method derives from empirical facts of procedural efficacy and reliability.[7]

---

7 It may, of course, be objected that on this model the source of such epistemic normativity remains quite unclear. The rules of method derive normative force from the goals toward which they are directed. But the hypothetical imperative analysis provides no basis

The problem remains, however, of the relation between method and truth. As discussed in 6.5, one cannot directly observe that use of the rules of scientific method leads to true scientific theories. The truth of the non-observational content of theories transcends empirical verification, hence cannot be established by direct observational means. It is at this point that appeal is to be made to the scientific realist argument that realism is the best explanation of the success of science. But where the success argument is usually employed to argue for the approximate truth of theories, I extend the argument to the truth-conduciveness of rules of method. I will now sketch the position of abductive realism, which seeks to bridge the gap between method and truth.

7.9 Abductive Realism

On the scientific realist picture that I propose, the relation between method and truth is not an analytic, conceptual relation, as the internal realist suggests, but a synthetic, empirical relation. It is a contingent relation between epistemic means and ends, which may be known in the *a posteriori* manner suggested by epistemic naturalism. But the attempt to combine a naturalistic account of epistemic warrant with the realist view of truth as the aim of science must face the problem that no empirical evidence may show directly or conclusively that use of a methodological rule yields theoretical truth. In the absence of direct or conclusive evidence, why should use of a rule of method be taken to conduce to truth?

This is where abductive realism enters the picture. In the absence of direct or conclusive evidence linking method to truth, the grounds for such a link may be at best abductive ones. More specifically, the realist claim that application of rules of method leads to progress toward truth rests on an inference to the best explanation of scientific success. What best explains why scientific theories satisfy the rules of method is that they are close to truth.

Suppose, for example, that there is some theory which satisfies a broad range of rules of method to an extraordinarily high degree. The theory is supported by all available evidence. It successfully predicts a great many previously unknown and surprising novel facts. It unifies previously disparate domains. And it does all of this in a manner which maximizes simplicity and coherence. Clearly, any theory which so impressively satisfies the rules of scientific method is a highly successful theory indeed.

How is such success to be explained? Where a theory impressively satisfies a broad range of methodological rules, the best explanation of such success is that the theory provides an approximately true description of the way the world is. In light of such success, we may infer not only that the entities postulated by the theory exist in roughly the form stated by the theory, but that the underlying causal mechanisms and processes described by the theory really do bring about observable events in the general manner specified by the theory.

---

on which to evaluate the epistemic merits of any particular epistemic goal. This problem is resolved within the framework adumbrated here by treating truth as the ultimate goal of scientific inquiry from which the value of lower order epistemic goals is derived.

It is important to emphasize that the level of descriptive accuracy to which such an inference is committed is that of approximate truth only. While the precise natures of the postulated entities, mechanisms and processes may fail to be known either in detail or in their entirety, it may nevertheless be the case that such entities, processes and mechanisms really do exist, in a form which is close to that described by the theory. Given such approximate accuracy, it must also be emphasized that the theoretical description of the postulated entities, mechanisms and processes remains open to possible revision in the light of further inquiry.[8]

But my point is not simply that the best explanation of the success of a theory, as measured by satisfaction of methodological rules, is the approximate truth of the theory. The crucial point relates to the truth-conduciveness of methods rather than to the approximate truth of theory. Given the critical role played by the rules of method in the process of theory selection, the implications of the success of science for the approximate truth of theory apply with equal force to the rules of method themselves.

In particular, the rules of method are employed by scientists to eliminate theories that are unlikely to be true in favour of theories that are likely candidates for truth. Since the best explanation of satisfaction of rules of method is the approximate truth of theory, and since the rules of method play a critical role in arriving at such approximately true theories, it follows that use of the rules of method is responsible for arriving at theories that are approximately true. Given this, the best explanation of the role played by the rules of method is that the rules are employed in a rigorous selection process which eliminates false theories in favour of theories that are closer to the truth. That is, the rules of method are rules that 'screen for truth' – or, in other words, the rules of method reliably conduce to truth.

Such an abductive realist account of the truth-conduciveness of method has implications as well for the realist view of scientific progress as convergence on truth. Suppose there is a sequence of scientific theories which displays an increasingly high level of satisfaction of the rules of method. According to abductive realism, increased satisfaction of the rules of method is to be attributed to convergence on truth. Where a sequence of theories displays an increasingly high level of satisfaction of the rules of method, the best explanation is that the sequence of theories is advancing progressively closer to the truth.

Thus, on the view I propose, what best explains satisfaction of the rules of method is that the rules are truth-conducive, and what best explains increased satisfaction of such rules is convergence on truth. It is in this sense that I wish to claim that what is needed to bridge the gap between method and truth is an abductive argument to the best explanation of the success of science.[9]

---

8 In holding there to be a reasonably clear sense of 'approximate truth' which relates to the general ontological claims of theory, and does not require explication by means of a technical concept of verisimilitude or closeness to truth, I follow Ernan McMullin's discussion in his (1984, 35-6, and 1987, 59-60).

9 The strategy described here as 'abductive realism' is not without precedent in the epistemology of science. Broadly understood as inference to the best explanation of the

7.10 Realism or Sheer Luck?

It is, of course, a legitimate question why truth and approximate truth should play the role which I ascribe to them as the best explanation of satisfaction of rules of method. To demonstrate that satisfaction of the rules of method is best explained by the truth-conduciveness of such rules would require an exhaustive elimination of alternative explanations. I cannot undertake that task here. But it is instructive to consider a stark anti-realist alternative that is contrary to the abductive realist thesis that the rules of method are truth-conducive rules whose use promotes the discovery of truth about the world. By eliminating this alternative a large and particularly salient class of anti-realist alternatives may also be eliminated.

To generate such a contrary to abductive realism, let us consider the following scenario. Consider, as we did before, a scientific theory which impressively satisfies a great variety of methodological rules. The theory is descriptively accurate and well-confirmed by all observational tests. It predicts surprising novel facts in an accurate and reliable manner. It unifies phenomena from domains previously thought to contain disparate and unrelated phenomena. On top of all this, the theory is also maximally simple and coherent.

This time, however, let us also suppose that despite impressively satisfying all the methodological rules the theory is in fact *totally and utterly false at the non-observational level.* None of the unobservable entities, mechanisms or processes postulated by the theory exists. Moreover, the theory erroneously imposes unity on unrelated domains which in fact have nothing in common. In short, let us suppose that the theory satisfies all empirical and formal methodological constraints to a very high degree, yet at the level of the descriptive accuracy of its claims about the underlying nature of reality it is simply false.

If such a situation were to obtain, it would be sheer luck that the theory has any success at all. This may be seen most clearly in the case of predictive success, and, in particular, in the case of accurate and reliable prediction of previously unknown and otherwise entirely unexpected phenomena. Either predictive success of this kind is the result of sheer luck, or else there is some benevolent force whose action makes the theory's predictions turn out to be true despite the fact that the theoretical claims of the theory are completely false.

There are, I suppose, possible worlds in which lucky guesses are routinely rewarded with predictive success. But we do not live in such a world. Occasional guesses may succeed. But if a scientific theory reliably produces accurate predictions of novel facts, the best explanation of such predictive success is not that we live in a world that rewards luck. The best explanation is that the theory is at least an approximately correct description of the unobservable entities whose behaviour underlies the observed phenomena predicted by the theory. For this reason, we may conclude that satisfaction of methodological rules provides a reliable indication of advance on truth. The rules of method are a guide to the truth. They are a guide to the truth, not in the sense that truth consists in satisfaction of the rules of method,

---

success of science applied at the level of method, the strategy is employed by such authors as Boyd (1984, 58-9), Kornblith (1993a, 41-2), and Rescher (1977, 81ff).

but in the sense that a theory that satisfies such rules has a good chance of being at least approximately true. If a theory which satisfies the rules of method did not have a good chance of being at least approximately true, the satisfaction of the rules of method would be completely inexplicable.

Chapter 8

Why is it Rational to Believe Scientific Theories are True?

8.1 Introduction

In this chapter, I continue discussion of the relation between method and truth. The discussion proceeds in the context of an analysis of the philosophy of science of Alan Musgrave. Musgrave is one of the foremost contemporary defenders of scientific realism. He is also one of the leading exponents of Karl Popper's critical rationalist philosophy. My main focus in this chapter will be Musgrave's realism. But I will emphasize epistemological aspects of realism. This will lead me to address aspects of Musgrave's critical rationalism as well.

Musgrave is both a scientific realist and a commonsense realist. 'Scientific realism,' he says, 'is a form of realism' (1999, 132). And realism is committed to the commonsense realist belief 'that there is a real world outside of us and largely independent of us' (1999, 132). 'There is,' Musgrave adds, 'a continuity between common sense and science' (1999, 132). But while science may lead to occasional revision and refinement of common sense, 'it does not show that it is root-and-branch mistaken' (1999, 133; cf. 1996, 23). The real world postulated by common sense is the reality that science seeks to explain. This world does not depend on human belief or experience. Nor is it relative to conceptual scheme, theoretical background or mode of description (1999, 52, 173, 180ff).

For Musgrave, though, realism is not just a thesis about reality. It is also a thesis about truth. Musgrave takes the aim of science to be truth. He 'subscribe[s] to the old-fashioned idea that scientific realism ... says that the aim of a scientific inquiry is to discover the truth about the matter inquired into' (1996, 19; cf. 1999, 52). Scientific theories are taken at face-value as genuine assertions about the world, the truth or falsity of which depends on the way the world really is (1996, 26). Truth is understood in the classic correspondence sense defined by Tarski. A theory or statement is true just in case the world is the way it is said to be (1993, Ch. 14; 1996, 24; 1999, 165). This is a 'non-epistemic conception of truth' (1996, 28; cf. 1999, 186). Given the emphasis on correspondence between theory and reality, Musgrave's realism diverges from the tendency among some scientific realists to adopt ontological rather than truth-orientated versions of the doctrine. Musgrave dismisses such entity-realism as incoherent (1996, 20).[1]

1 Musgrave raises the following objection to entity realism: 'We are to believe in scientific entities ... without thinking true any theory about those entities ... This is incoherent. To believe in an entity, while believing nothing further about that entity, is to believe nothing.

Musgrave's realism has an epistemological dimension as well. For Musgrave, methodological considerations play a prominent role in the appraisal and acceptance of scientific theories. While a variety of methodological rules figures in Musgrave's writings, there is some tendency on his part to emphasize the testing and falsification of theories.[2] The attempt to falsify theories is the basis of the critical method in science. And criticism is the heart of rationality. A critical discussion may provide 'the best reason there is for believing (tentatively) that a hypothesis is true' (1999, 324). If a theory 'best withstands criticism then it is reasonable for scientists to believe that theory and to use it in practical applications' (1999, 325). Such belief must remain tentative, however, for Musgrave is a fallibilist who eschews the search for epistemic certainty in science and everyday affairs (cf. 1993, Ch. 15; 1999, 194ff, 341-3).

But matters of method and rationality are separate matters from those of reality and truth. This is especially the case from the perspective of realism. In the first place, to believe that the world is a given way does not mean that the world is that way. Nor does it make the world that way. Reality is not subject to determination by human thought. This remains the case even if the belief that the world is a given way is a belief that is rationally justified. For one may rationally believe what is false. The point applies with equal force to scientific theories certified by the rules of scientific method. A theory that is certified by the rules of method is not thereby shown to be true. A theory which satisfies methodological rules may yet be false. Nor need a theory that satisfies methodological rules be accepted as true. The methods of science are not the exclusive domain of realism. They may serve aims other than the realist aim of truth. Satisfaction of the rules of method might indicate empirical adequacy or pragmatic reliability, rather than truth.

An explanation is therefore required on the part of the realist of why certification by method provides warrant with respect to truth. I will refer to the need to provide

---

I tell you that I believe in hobgoblins (believe that the term 'hobgoblin' is a referring term). So, you reply, you think there are little people who creep into houses at night and do the housework. Oh no, say I, I do not believe that hobgoblins do that. Actually, I have no beliefs at all about what hobgoblins do or what they are like. I just believe *in* them' (1996, 20). Musgrave's point is that it is not possible to believe *in* the existence of some entity without having at least some beliefs *about* the entity. This is a crucial point to be made in relation to entity realism. But it does not entirely dispose of the doctrine. For, as Musgrave notes, entity realists may adopt a less extreme position according to which some low-level theoretical beliefs may be true of the theoretical entities.

2 Since Musgrave often writes within the context of falsificationist philosophy of science, an emphasis on such issues as corroboration, independent testability, *ad hoc*ness and predictive novelty is perhaps understandable. However, within the context of scientific realism, Musgrave places special emphasis on the role of novel predictions, arguing that the success argument for scientific realism should be restricted to theories which correctly predict facts not employed in the construction of the theory (cf. Musgrave, 1999, 55-7, 119, Ch. 12). Other methodological criteria, such as simplicity or unity, also receive favourable mention (cf. 1999, 111-2, 247ff). Thus, despite the emphasis on falsification, Musgrave allows that the methodology of science consists of a plurality of methodological rules (cf. 1999, 226-7, 250, fn. 291).

such an explanation as *the problem of method and truth*. As a realist who holds that it may be rational to believe a theory which has been subjected to critical scrutiny in accordance with the rules of method, the problem of method and truth is one that Musgrave must address. That is, he must confront the question of why it is rational to believe theories certified by the methods of science to be true, or close to the truth.[3] In this chapter, I will explore his response to the problem.

I will illustrate the problem of method and truth in section 8.2 by means of the examples of Lakatos's 'plea for a whiff of inductivism' and the internal realist conception of truth of Putnam and Ellis. In section 8.3, I will turn to Musgrave's approach to the problem of method and truth, where I will consider his treatment of inference to best explanation and critical rationalism. In section 8.4, I will explore a naturalistic approach to the problem which sets the issue within a broader metaphysical framework. Finally, in section 8.5, I shall offer some suggestions as to how Musgrave might put metaphysical aspects of his position to epistemological use.

8.2 The Problem of Method and Truth

Scientific realism enforces a sharp divide between method and truth. On the one hand, scientific method consists of a set of rules and procedures which govern experimental practice and inform the appraisal of scientific theories. A scientist whose acceptance of a theory or result complies with the rules and procedures of method is rationally justified in accepting the theory or result. On the other hand, truth consists in a relation of correspondence between a statement and extralinguistic reality. The relation of correspondence between statement and reality is a relation that may obtain whether or not one has methodologically warranted grounds for believing it to obtain. Indeed, it is a relation that may obtain whether or not the statement is believed to be true. Truth, in the correspondence sense, is a non-epistemic relation, which is not defined in terms of method or rational justification.

Given the separation of method and truth, the question arises of the relation between them. What bearing does method have on truth? Why should use of method lead to theories that are either true or approximately true? This is the problem of method and truth. To illustrate it, I will now turn to Lakatos's 'whiff of inductivism' and the internalist conception of truth of Putnam and Ellis.

Lakatos's Plea for a 'Whiff of Inductivism'

The problem of method and truth may be illustrated within the context of Popper's philosophy of science by means of the connection between corroboration and verisimilitude. For Popper, a theory is corroborated by successful performance

3 The problem of method and truth is not restricted to truth-orientated forms of realism. For the entity realist must face exactly the same challenge of explaining why use of the methods of science leads to knowledge of the way the world is. The problem is the general one of explaining how a methodological procedure conduces to knowledge of an objective reality.

in an empirical test of a prediction made by the theory. The theory receives high corroboration if it passes a range of such tests, especially ones which comprise severe tests of the theory. By contrast, the concept of verisimilitude is a measure of the truth-content relative to the falsity content of a theory, which Popper proposes as an analysis of the idea that one theory may contain more truth than another. One theory has greater verisimilitude than another if it has greater truth-content relative to falsity content than the other.

The question is whether there is any reason to believe that a theory with a higher degree of corroboration than another should also enjoy a higher degree of verisimilitude than the other. In other words, is corroboration an indication of verisimilitude?

In his contribution to *The Philosophy of Karl Popper*, edited by P.A. Schilpp, Imre Lakatos expresses the concern that Popper's 'fallibilism is nothing more than scepticism together with a eulogy of the game of science' (1974, 257). Lakatos's concern is precisely that, as a fallibilist and anti-inductivist, Popper is not prepared to:

> say unequivocally that the positive appraisals in his scientific game may be seen as a – conjectural – sign of the growth of conjectural knowledge; that corroboration is a *synthetic* – albeit conjectural – measure of verisimilitude. (1974, 256)

Nor may Popper assert that high corroboration provides any positive reason to believe that a theory is close to the truth.

In order to address this concern, Lakatos enters a plea for a 'whiff of inductivism' to the effect that Popper's methodology be supplemented with a 'synthetic inductive principle' (1974, 254-7, 260). Such a principle would connect corroboration with verisimilitude by treating the former as a 'sign' or 'measure' of the latter (1974, 254, 256). Only in this way, Lakatos argues, can the methodological concept of corroboration and the 'logico-metaphysical' notion of verisimilitude be combined into a properly epistemological theory of the growth of scientific knowledge.

In his reply to Lakatos, Popper does not explicitly address the plea for a synthetic inductive principle. He does, however, allow that corroboration serves as an 'indication' of verisimilitude in the sense that 'we may guess that the better corroborated theory is also one that is nearer to the truth' (Popper, 1974a, 1011). But he denies that corroboration is to be understood as in any sense a measure of verisimilitude.[4]

There is one point in the Schilpp volume, though, where Popper does seem to concede a 'whiff of inductivism'. In his reply to A. J. Ayer, Popper explains the importance of the notion of verisimilitude:

> ... there is a probabilistic though typically noninductivist argument which is invalid if it is used to establish the probability of a theory's being true, but which becomes valid (though essentially nonnumerical) if we replace truth by verisimilitude. The argument can be used only by realists who do not only assume that there is a real world but also that this world is by and large more similar to the way modern theories describe it than to the

4 See also Popper (1972, 103). For related discussion, see Newton-Smith (1981, 67-70).

> way superseded theories describe it. On this basis we can argue that it would be a highly improbable coincidence if a theory like Einstein's could correctly predict very precise measurements not predicted by its predecessors unless there is "some truth" in it. (Popper, 1974b, 1192-3, fn. 165b)

Popper goes on to remark that 'there may be a "whiff of inductivism" here', which 'enters with the vague realist assumption that reality, though unknown, is in some respects similar to what science tells us' (1974b, 1193).

It is unclear why Popper fails to make this concession in the context of his response to Lakatos. In any event, the assumption of a real world that is 'by and large similar to the way modern theories describe it' would appear to be a metaphysical assumption of the very kind that Lakatos proposes. If there is a real world which contains the entities and laws which science tells us that it contains, then this fact is itself the explanation of why contemporary theories which say that there are such entities and laws receive high corroboration. For if the world contains things which do what a theory says they do, then *that* is why what the theory says about those things is true. But such an explanation may only be provided on the assumption that theories which succeed in the manner indicated by high corroboration are close to the truth.[5]

As I will attempt to show in sections 8.3 and 8.4, it is precisely such an appeal to metaphysics that is lacking from the epistemology of Musgrave's realism. In this respect, Musgrave seems to side with Popper against Lakatos in resisting the call for a metaphysical inductive principle. But, as I will attempt to show, to defend the epistemological basis of realism, the realist must put the world to good use.

Putnam on the Ideal Limit of Inquiry

As we have seen, Lakatos proposes to bridge the gap between method and truth by means of a 'synthetic inductive principle'. An alternative approach is to close the gap in an analytic manner by defining truth in terms of method. This is the path of internal realism (e.g., Putnam, 1978, 1981; Ellis, 1990). In this section, I will briefly explore this path before indicating why it is not one that can be taken by the realist. Since Musgrave has forcefully argued for this conclusion, I will draw on his work in showing that realism cannot go down the internalist path.

In his (1978), Hilary Putnam notes that according to the position which he describes as 'metaphysical realism', truth is 'radically non-epistemic' (1978, 125).[6]

---

5 This is not to say that the connection between the approximate truth or verisimilitude of a theory and its empirical success is unproblematic. In fact, it cannot be assumed that a theory with a high degree of approximate truth will be successful. For example, many of its observational claims might be false even though it contains a great deal of true theoretical claims (cf. Laudan, 1984, 118). But the present point is not that there is an unproblematic connection between approximate truth and success. Rather, the point is that Popper appears to make a metaphysical assumption about the nature of reality, on the basis of which some non-analytic relation between verisimilitude and corroboration might be shown to obtain.

6 Putnam's characterization of metaphysical realism contains elements which may not be acceptable to all realists. In his 'Metaphysical Realism versus Word-Magic' (2001),

For metaphysical realism, truth is a semantic relation of correspondence between linguistic items and entities in the external world. Such a concept of truth is defined independently of epistemic factors, such as evidence, confirmation or simplicity.

Putnam illustrates the non-epistemic nature of metaphysical realist truth with the example of the ideal theory which would ultimately result if science were pursued to the ideal limit of inquiry. Such a theory would maximally satisfy all methodological constraints. Putnam says the ideal theory would be:

> ... complete, consistent ... predict correctly all observation sentences ... meet whatever 'operational constraints' there are ... be 'beautiful', 'simple', 'plausible', etc ... (1978, 125)

Given the non-epistemic nature of truth, however, it is possible that even such an ideal theory might be false. For while it might be extraordinarily unlikely for the ideal theory to be false, the fact that it maximally satisfies all methodological constraints does not entail that it is true.

Putnam rejects both metaphysical realism and the non-epistemic conception of truth.[7] He proposes instead an internal realist stance on which truth is understood in epistemic terms as an idealized form of rational justification:

> 'Truth', in an internalist view, is some sort of (idealized) rational acceptability – some sort of ideal coherence of our beliefs with each other and with our experience as those experiences are themselves represented in our belief system ... (1981, 49-50)

The internalist conception of truth differs from the metaphysical realist conception on two counts. First, it is an epistemic conception of truth which takes truth to be a form of rational acceptability. Second, because truth is idealized rational acceptability, the epistemically ideal theory produced at the ideal limit of scientific inquiry must necessarily be true.

Musgrave argues that realists should not uncritically accept the idea of a mind-independent reality, since there is a range of mind-dependent objects (e.g. artifacts) about which one should be thoroughly realist. Musgrave also objects to the idea that there is a 'fixed totality' of mind-independent objects, since what objects there are depends on a prior specification of what sort of object is in question. (For further discussion, see 2.3.)

7 Putnam presents a number of objections to metaphysical realism. One is that truth is not radically non-epistemic because the ideal theory cannot possibly be mistaken. This objection rests on his well-known model-theoretic argument against realism that since every consistent theory has at least one model, the ideal theory (which is stipulated to be consistent) must be true (Putnam, 1978, 125-6). A second objection is that in order to describe the position of metaphysical realism it must be possible to adopt a God's eye point of view. But it is impossible to remove ourselves from our limited human perspective to adopt the external viewpoint of such an omniscient being (Putnam, 1981, 50). A third objection is that metaphysical realism opens the door to the possibility of radical scepticism, since it allows the possibility of massive illusion (e.g., evil demons, brains in vats). But such radical sceptical scenarios are not in fact possible scenarios. Hence, metaphysical realism is mistaken because it allows the possibility of such scenarios (Putnam, 1981, 15).

The internal realist conception of truth provides a clear example of one way to deal with the problem of method and truth. The internalist closes the gap between method and truth by setting up an analytic or conceptual relation between method and truth. If truth just is a form of rational justification, then a theory which satisfies methodological rules of theory-acceptance is to be accepted as true, or nearly so. For that is what it is to be true. Equally, a theory which better satisfies methodological rules than a predecessor thereby displays a higher degree of truth, since increased satisfaction of such rules constitutes increase of truth.

Such an analytic resolution of the problem of method and truth is not, however, one that is open to the scientific realist. For, as Musgrave has argued, the internalist conception of truth leads to an idealist metaphysics that is unacceptable to realists. In his paper, 'The T-Scheme Plus Epistemic Truth Equals Idealism' (1999, Ch. 10; cf. 1996, 30), Musgrave argues that epistemic theories of truth, such as internal realism, entail the dependence of reality upon belief.[8] According to Musgrave, 'the general form of an epistemic truth theory' is as follows:

> Necessarily, S is true if and only if S satisfies epistemic condition E. (1999, 188)

To obtain a particular epistemic theory of truth from this general form, it suffices to replace the epistemic condition (E) with the preferred epistemic condition of the relevant truth theory.

Musgrave employs the example of Brian Ellis's evaluative theory of truth, which is a form of internal realism closely related to Putnam's. According to Ellis, truth is what it is epistemically right to believe. So we have:

> Necessarily, S is true if and only if it is epistemically right to believe S.

Now, given the T-scheme:

> (T) S is true if and only if P,

the evaluative theory of truth entails that:

> Necessarily, P if and only if it is epistemically right to believe S.

Thus, to take a particular example:

> (ET) Electrons exist if and only if it is right to believe that electrons exist. (1999, 189)

But, surely, Musgrave points out, (ET) might be false. There might be no electrons even though 'our best methods optimally pursued ... lead us to think electrons exist' (1999, 189). The only way for (ET) to be true is for the world to depend on our methods of inquiry or our theories in idealist fashion. In this case, electrons would exist if that is what our methods of inquiry and theories lead us to believe. But that is evidently not something that a realist can accept.

8 For related analysis, see Devitt and Sterelny (1987, 196).

The Problem Restated

Lakatos's plea for a 'whiff of inductivism' and Putnam and Ellis's internalist conception of truth represent two different approaches to the problem of method and truth. The question is why we should suppose that the rules of method have any positive bearing on truth. The response proposed by Putnam and Ellis is to define truth in terms of method. But such a response is unavailable to the realist who takes truth to be non-epistemic, as Musgrave does. The other response which we have seen is to appeal to a synthetic metaphysical principle in the manner suggested by Lakatos with his 'plea for a whiff of inductivism'. But this response appears not to be the response favoured by Musgrave, as we will now see.

8.3 Musgrave on Method and Truth

As a scientific realist, Musgrave adheres to the view that it may be rational to believe that a scientific theory is true. A theory which passes critical scrutiny by means of the rules of scientific method may be accepted as true, where truth is understood in the non-epistemic sense of the realist. The question is why it is rational to believe that a theory which satisfies the rules of method is true. If truth is non-epistemic, then what does method have to do with it?

In this section, I will consider two answers that have been proposed by Musgrave. The first involves the idea that it is reasonable to believe the best explanation of a fact. The second is that it is rational to believe the hypothesis which best survives criticism. As we will see, neither approach succeeds in showing why it is rational to believe a theory to be true.

'The Ultimate Argument for Scientific Realism'

The standard argument for scientific realism is the so-called 'success argument', or, as Musgrave calls it, 'the Ultimate Argument'.[9] According to scientific realism, the entities postulated by mature scientific theories by and large exist, and the claims that theories make about those entities are by and large true, or close to the truth. Such a realist account of the relation between theories and the entities they postulate provides a compelling explanation of the empirical success of science. For if the entities postulated by a theory exist, and what the theory says about the entities is true, then it is no surprise that the theory should meet with empirical success. By contrast, any anti-realist philosophy which rejects the realist view of the relation between theories and the entities they postulate must render the success of science an inexplicable miracle. But to say that the success of science is a miracle is to fail to provide an adequate explanation of such success. Since realism provides a compelling explanation of success, and anti-realism fails to provide an adequate explanation, realism is evidently the best explanation of the success of science.

---

9 The name, 'the ultimate argument', is due to van Fraassen (1980, 39), who is one of the targets of Musgrave (1988).

In his paper, 'The Ultimate Argument for Scientific Realism' (1988, 232-9), Musgrave presents an analysis of the success argument.[10] It is standard practice to construe the success argument as an inference to the best explanation. In line with this practice, Musgrave also construes the argument as an inference to the best explanation. However, in a novel departure, Musgrave argues that application of the success argument is to be restricted to theories which successfully predict novel facts. He formulates the argument as an epistemic argument to the effect that it is reasonable to accept realism, rather than to the effect that realism is true. He further stipulates that in order to be acceptable, the best explanation must satisfy minimal conditions of explanatory adequacy. Otherwise, it would not be reasonable to accept the best explanation as true.

Opinion is divided over the nature of inference to the best explanation. Some take it to be a form of inductive inference. Others take it to be a *sui generis* form of inference that is more fundamental than induction. Perhaps the most novel feature of Musgrave's analysis of the success argument is his suggestion that inference to the best explanation may be formulated as a deductive inference.

Musgrave proposes that inference to the best explanation be construed in deductive form as follows:

> It is reasonable to accept a *satisfactory* explanation of any fact, which is also the *best* available explanation of that fact, as true.
>
> *F* is a fact.
>
> Hypothesis *H* explains *F*.
>
> No available competing hypothesis explains *F* as well as *H* does.
>
> Therefore, it is reasonable to accept *H* as true.
>
> (Musgrave, 1988, 239)

He then comments that 'the Ultimate Argument for scientific realism ... is an inference to the best explanation':

> The fact to be explained is the (novel) predictive success of science. And the claim is that realism ... *explains* this fact, explains it *satisfactorily*, and explains it *better* than any non-realist philosophy of science. And the conclusion is that it is reasonable to accept scientific realism ... as true. (Musgrave, 1988, 239)

On such a construal, the success argument is a valid deductive argument. The fact to be explained is the novel predictive success of science. The conclusion of the argument is an epistemic conclusion to the effect that it is rational to believe realism to be true. For realism is the best explanation of predictive success. The conclusion

10 I refer here to the original version of Musgrave's article in Nola (1988). The paper is reprinted in Musgrave (1999). However, the section of the article on inference to best explanation, which is of central relevance to scientific realism, has been removed. It appears, instead, in the context of a discussion of psychologism (1999, 284-5).

depends crucially on the epistemic principle that it is reasonable to accept the best satisfactory explanation of a fact as true, which figures as the initial premise of the argument.

Musgrave's analysis of the success argument is an important advance in a number of respects. The emphasis on predictive novelty is important because it may be employed to eliminate a number of historical counterexamples which have been proposed to the success argument.[11] Musgrave's formulation of the success argument in epistemic terms makes clear that the argument must play a pivotal role in response to anti-realist critics who object to scientific realism on epistemological grounds. His emphasis on minimal conditions of explanatory adequacy is crucial, since it excludes the possibility that the best available explanation fails to be a satisfactory explanation. Finally, the explicit use of the epistemic principle in the argument makes evident the extent to which the success argument depends on the assumption of the epistemic importance of explanation.

Despite initial appearances, however, Musgrave's analysis of the success argument provides little assistance in relation to the problem of method and truth. To see this, let us further examine the notion of a *best* explanation. On what might the judgement that a theory is the best explanation be based? Musgrave does not elaborate. But it seems reasonably clear that the assessment of the explanatory merit of a scientific theory will depend upon methodological criteria of theory appraisal. Relevant criteria will include considerations of explanatory strength and unification, as well as simplicity, coherence and fit with background knowledge. But since truth is understood by Musgrave in the non-epistemic realist sense, it is unclear why theories which satisfy such methodological criteria should be accepted as true.

The question is why it is reasonable to accept the best explanation as true. Might it not be equally reasonable to accept the best explanation as empirically adequate, useful for practical purposes, or even true in some non-realist sense? Nothing Musgrave says in support of the principle that it is reasonable to accept the best explanation as true shows that the anti-realist might not accept an anti-realist analogue of the principle. Nor does Musgrave provide an explanation of why it is reasonable to accept the best explanation as true.

It might, however, be thought that the issue is not whether the best explanation is to be accepted as true. Rather, the issue is whether realism is the best explanation. Musgrave addresses this issue in the pages that follow his analysis of the success argument (1988, 240-4). He considers a range of anti-realist explanations of predictive success, and argues that all provide inferior explanations to the realist explanation. On the assumption that realism has been shown to be a superior explanation to anti-realism, it might therefore appear that realism is to be accepted as true.

But this only succeeds in pushing the problem back another level. Even if it is granted that realism is the best explanation of the success of science, it does not follow that it is to be accepted as true. There are other possible modes of acceptance

11 It is a major weakness of earlier formulations of the success argument that the notion of success is imprecisely defined. If success is left overly vague, the success argument is vulnerable to historical counterexamples, such as those presented by Laudan of theories which attained a degree of success but were false and/or non-referential (Laudan, 1984, Ch. 5).

available at this level, apart from acceptance as true. For example, one might simply agree that realism is the best explanation without proceeding to accept it as true. Alternatively, one might merely accept realism as if it were true. Or realism might be accepted as true, but truth might be understood in some non-realist sense. Nothing about best explanation, as such, clearly precludes such alternative forms of acceptance.

In sum, to show that a theory is the best explanation of a fact does not entail that the theory is to be accepted as true. Given this, Musgrave's analysis of the success argument in terms of an epistemic principle of best explanation does not succeed in showing why it is rational to accept a theory as true. It does not, in other words, provide a response to the problem of method and truth.

Critical Rationalism

I turn now to a second context in which Musgrave addresses issues which relate to the problem of method and truth. In his treatment of Popper's solution of the problem of induction, Musgrave proposes a critical rationalist account of scientific theory acceptance (Musgrave, 1999, Ch. 16). I will now consider the implications of Musgrave's critical rationalism with respect to the problem of method and truth.

Popper's philosophy of science is sometimes described as 'negativist' (cf. Lakatos, 1974, 258). In an attempt to solve Hume's problem of induction, Popper dismisses induction as a myth. Instead of offering a positive justification of induction, Popper argues that the attempted falsification of a theory may provide rational grounds for tentative acceptance of the theory. It is possible neither to prove that a theory is true nor to provide inductive support for the theory. However, if a theory has survived rigorous empirical tests, then it may be rational to tentatively accept the theory.

Since Popper denies that there may be any grounds which provide positive support for a theory, the question arises of how his claim that it may be rational to accept a theory is to be understood. To address this question, it is necessary to introduce a distinction between Popper's critical rationalist account of rationality and the traditional *justificationist* conception of rationality to which Popper's account is opposed. Perhaps what most fundamentally characterizes Popper's account of rationality is his outright dismissal of the justificationist conception of rationality.

The justificationist conception of rationality is the conception of rationality that underlies most traditional and contemporary thinking about rational belief. According to justificationism, in order to have a rational belief the belief itself must be rationally justified. There must be reasons which provide support for the belief.

Musgrave characterizes justificationism by means of the following principle:

> (J) A's believing that P is reasonable if and only if A can justify P, that is, give a conclusive or inconclusive reason for P, that is, establish that P is true or probable. (Musgrave, 1999, 321)

As this formulation of justificationism makes clear, reasons may either be conclusive or inconclusive. Conclusive reasons are reasons which show that a belief is true.

Inconclusive ones merely show it to be likely or probable. In either case, rational belief requires there to be reasons which support the belief itself.

By contrast with justificationism, critical rationalists deny that there may be reasons for a belief or theory. But this does not mean that there is no rationality. On the contrary, as Popper remarked, 'there is nothing more "rational" than the method of critical discussion, which is the method of science' (1972, 27). Criticism, rather than justification, is the key to rationality.

Accordingly, Musgrave offers the following principle as formulation of critical rationalism:

> (CR) It is reasonable to believe that P (at time t) if and only if P is that hypothesis which has (at time t) best withstood serious criticism. (Musgrave, 1999, 324)

In other words, if a hypothesis is subjected to serious criticism and survives, while alternative hypotheses do not, there is good reason to accept the hypothesis which stands up to criticism in favour of those which succumb to it. By contrast with justificationism, such a conception of rationality does not involve good reasons for a hypothesis. It is belief in the hypothesis, rather than the hypothesis itself, for which there may be good reason. Critical rationalism alters the locus of rationality. 'It is', Musgrave explains, 'acts of belief (actions of believing?) that are reasonable or rational, not the things we believe, belief-contents, propositions, theories, or whatever' (1999, 322).

On Musgrave's analysis of critical rationalism, it is rational to believe 'the theory which best survives critical scrutiny' (1999, 330). To believe a theory is to believe that it is true (cf. 1999, 321, 326). And the method of criticism is the method of science. The critical rationalist account of theory acceptance is therefore of clear relevance to the problem of method and truth. For the critical rationalist asserts that survival of critical scrutiny provides the basis for rational belief in the truth of scientific theories.

But what is it for the method of criticism to be the method of science? As earlier noted, within the context of a Popperian falsificationist theory of method, the primary means of criticism is the attempt to falsify a theory by rigorous empirical test. Within a strictly falsificationist framework, it is possible to criticize a theory in a variety of ways. A theory may entail a false prediction or it may be unfalsifiable. It might predict no novel facts, be poorly corroborated, or be *ad hoc*. But there is no need for the method of criticism to be restricted to strictly falsificationist resources. A theory might also be criticized on grounds which have no immediate connection with empirical falsification as such. For example, a theory might lack coherence, be overly complex, have limited explanatory scope, or be inelegant.

A variety of methodological considerations may therefore play a role in the critical method. But it remains to be asked how the critical method warrants belief in the truth of a theory. By itself, the rejection of justificationism does not suffice to resolve the problem of method and truth. If truth is non-epistemic, and the critical method is the basis of theory acceptance, the connection between method and belief in the truth is left entirely unexplained.

It would be misguided to suppose that survival of criticism provides positive support for a theory. For the critical rationalist, survival of rigorous test or other attempts to criticize a theory does not lend positive support to a theory. To assume that criticism yields positive support is to assume a justificationist conception of rationality. But, for the critical rationalist, survival of criticism does not prove that a theory is true, nor does it render the theory more likely to be true. It does not provide any positive justification for the theory at all. Rather, survival of criticism provides one with a basis to tentatively believe in the truth of a theory, as opposed to alternative theories which have been exposed to criticism and failed to survive.

The trouble is that nothing has been done to secure belief in truth as the unique mode of theory acceptance. It is possible to agree with the critical rationalist conception of scientific inquiry, but to deny that theories are to be accepted as true. To take but one example, it would be perfectly consistent for an anti-realist to endorse the critical method while at the same time embracing a constructive empiricist view of theory acceptance along the lines of Bas van Fraassen.[12] On such an account, it would be rational to accept a theory which best withstands critical scrutiny. But the theory is to be accepted as empirically adequate, rather than true. That is, it is to be accepted as true at the observational level, without commitment to the truth of its non-observational content.

Nothing about the critical method entails that a theory which survives criticism is to be accepted as true. Critical rationalists are fallibilists. As such, critical rationalists themselves insist that a theory which survives rigorous empirical test may fail to be true. But, if it does not follow from survival of criticism that a theory is true, then neither does it follow that the theory is to be accepted as true. There is nothing about the notion of criticism as such which requires one to believe that a theory which survives criticism is true.

Musgrave introduces a modification of critical rationalism which may seem to go some way toward disarming this objection. The modification relates to the 'epistemic primacy' of perception (1999, 342). Perception is the source of the empirical evidence which is employed to test our theories. But on what basis are perceptual reports accepted? In ordinary circumstances, perceptual reports are not accepted as the result of test. Rather, they are accepted at face value. Perception is only subjected to test when something goes wrong. As Musgrave notes, 'only when we have some specific reason to suspect perceptual error do we "check out" a perceptual belief' (1999, 342). But if it may be rational to accept a perceptual report which has not been subjected to test, then survival of criticism cannot be necessary for rational belief.

This point requires that critical rationalism be amended. For if it may be rational to accept a perceptual belief without submitting it to test, then it may be rational

12 Indeed, van Fraassen comes close to such a position when he remarks that 'the success of current scientific theories is no miracle. It is not even surprising to the scientific (Darwinist) mind. For any scientific theory is born into a life of fierce competition, a jungle red in tooth and claw' (1980, 40). Of course, this remark is made in the context of van Fraassen's discussion of the realist's success argument. But the talk of fierce competition suggests that van Fraassen approaches the question of theory acceptance with a decidedly Popperian cast of mind.

to accept such a belief without it having survived criticism. Musgrave, therefore, introduces a distinction between perceptual and non-perceptual beliefs:

> A non-perceptual belief is reasonable if it has best withstood criticism – a perceptual belief is reasonable if it has not failed to withstand criticism. The latter is just the commonsense view 'Trust your senses unless you have a specific reason not to'. (1999, 342)

On the modified version of critical rationalism to which this distinction gives rise, rational theory acceptance requires survival of criticism. But perceptual belief is rational provided only that no problem has so far arisen with respect to the perception on which it is based.

But even if the primacy of perception is granted, this does not affect the objection. It may simply be conceded that perception provides a *prima facie* rationale for the acceptance of a perceptual report. No such rationale is thereby provided for theory acceptance. This is particularly apparent in light of Musgrave's epistemic distinction between perceptual and non-perceptual belief. The primacy of perception specifically relates to perceptual belief. Nothing follows from the primacy of perception with respect to the rationality of non-perceptual belief. If the primacy of perception is to be of any relevance to theory acceptance, then an additional assumption is required which extends the primacy of perception to the non-perceptual realm.

The point may be illustrated by means of the earlier example of the constructive empiricist version of critical rationalism. Such a constructive empiricist accepts the critical rationalist account of theory acceptance with the qualification that theories which survive criticism are to be accepted as empirically adequate. It is entirely consistent with such a position to grant the epistemic primacy of perception, and to agree that perception provides a *prima facie* rationale for perceptual belief. But the primacy of perception only entails that perceptual beliefs be accepted as true. It does not extend to the level of theory. Hence, the constructive empiricist may restrict theory acceptance to empirical adequacy.

Thus, even if the primacy of perception is granted, it does not follow that theories which pass critical scrutiny need be accepted as true. Given this, and the earlier point that survival of critical scrutiny does not entail belief in the truth of a theory, I conclude that the critical rationalist position presented by Musgrave does not resolve the problem of method and truth. It remains to be shown why use of the critical method provides any reason to believe that a theory is true.

Epistemic versus Metaphysical Principles

We have now considered two approaches proposed by Musgrave which are of relevance to the problem of method and truth. Both of the approaches are based on epistemic principles of rational belief. As such, both of the approaches proposed by Musgrave contrast with the approaches to the problem of method and truth canvassed in section 8.2.

In section 8.2, we first considered Lakatos's 'plea for a whiff of inductivism' that Popper's methodology be supplemented by a metaphysical principle which connects corroboration with verisimilitude. Such a principle would consist of a substantive synthetic claim about the world in the light of which corroboration is revealed to be an indication of verisimilitude. By contrast, Musgrave's epistemic principles say nothing about the world. Instead, they specify conditions under which it may be rational to believe a proposition or hypothesis to be true.

After considering Lakatos's proposal, we examined the analytic approach to the problem of method and truth that is due to internal realism. The internalist identifies truth with satisfaction of methodological criteria. Given such an identification, it may be rational to believe that a theory which satisfies methodological criteria is true. For that is what it is to be true.

By contrast with internal realism, Musgrave is a realist for whom truth is a non-epistemic correspondence relation. As such, Musgrave must reject the analytic approach on two counts. As a realist, he must reject the internalist conception of truth because of the idealism to which it leads. And as an advocate of a non-epistemic conception of truth, he must reject the internalist identification of truth with satisfaction of epistemic criteria.

But while it is clear that Musgrave must reject the analytic approach, it is not entirely clear why he rejects metaphysical principles in favour of epistemic principles of rational belief. It may be that Musgrave rejects metaphysical principles because he takes them to be inductive principles of the uniformity of nature of a kind that Hume showed to be unjustified (cf. Musgrave, 1993, 157ff). It may be that he takes the rejection of justificationism to entail the rejection of metaphysical principles (cf. Musgrave, 1999, 327). It may be that he takes there to be no need for metaphysical principles over and above scientific theories which may be accepted on critical rationalist grounds (Musgrave, 1999, 328-9). Or it may be that he takes such principles to rest on an anthropocentric metaphysics (Musgrave, 1999, 283, 285).

Whatever Musgrave's exact reason for rejecting metaphysical principles may be, I shall now attempt to show that such principles are necessary in order to solve the problem of method and truth. The truth of an empirical claim about the world depends upon the way that the world in fact is. In order to show that use of an epistemic method leads to such truth about the world, it is necessary to say something about the world. Otherwise, no connection is made between method and truth. In short, the problem of method and truth is at least partly one of metaphysics.[13]

8.4 Metaphysics and Naturalism

To illustrate the relevance of metaphysical considerations to the problem of method and truth, I will now examine two examples of the epistemological application of metaphysical considerations. The first case is that of Nicholas Rescher's *methodological pragmatism*. The second is Hilary Kornblith's grounding of inductive inference in natural kinds.

13 Musgrave is not completely dismissive of metaphysical principles. Against those who treat laws and theories as inference licenses, Musgrave claims that they may be under the influence of a positivistic bias against metaphysics (1999, 283). Moreover, he notes against positivism that metaphysical principles of theory construction may play a significant role in science and may even be subject to rational appraisal (1999, 309).

Rescher's Methodological Pragmatism

For the classical pragmatist, a true proposition is one the acceptance of which leads to practical success. Rescher refers to such pragmatism as *thesis pragmatism*, since it relates to specific propositions or theses. He rejects the pragmatist view of truth in favour of a correspondence conception. Instead of thesis pragmatism, he proposes a *methodological pragmatism*, which applies the criterion of practical success at the level of the methods of inquiry. The rules of method are to be evaluated in the manner of instruments in terms of their success in practical application. If a rule reliably performs the function for which it is designed, it thereby receives pragmatic justification (Rescher, 1977, 3-4). By contrast, individual claims are not practically justified, but receive indirect support from the methods by which they are certified (Rescher, 1977, 71-2).

For Rescher, pragmatically warranted methods of inquiry are to be regarded as 'truth-indicative' (1977, 83).[14] A proposition which satisfies a rule of method is therefore to be accepted as true. Thus, while truth and utility are distinct at the level of propositions, Rescher takes pragmatic success to have a bearing on truth at the level of method. Because Rescher takes certification by rules of method to warrant acceptance as true, his methodological pragmatism is therefore of relevance to the problem of method and truth. The question is why practical justification of method should be taken to be truth-indicative. The answer, as we shall now see, turns on metaphysical considerations.

In order to explain how practical success relates to truth, Rescher places the use of method within a broader metaphysical setting. This is characterized by the following principles which relate to human agency, the community of inquirers and the nature of reality (1977, 84-9). *Activism:* our survival and welfare require action on our part; since we act on the basis of beliefs, our beliefs are of practical relevance. *Reasonableness*: belief guides action in a way that coordinates action with beliefs and needs. *Interactionism*: our active intervention in the world produces outcomes which may either satisfy or frustrate our intentions. *Purposive constancy*: to establish the reliability of a method, inquirers must employ the same method for the same purpose. *Uniformity of nature*: continued use of a method depends upon the underlying constancy of nature and the conditions of application of the method. *Nonconspiratorial nature of reality*: nature is indifferent to our beliefs and needs, neither conspiring for nor against belief-based actions.

Against this metaphysical backdrop, Rescher argues that a method of inquiry whose use systematically meets with success is to be seen as truth-indicative. False

14 Rescher's expression 'truth-indicative' may seem to suggest that a proposition that satisfies a methodological rule is thereby definitively shown to be true. Indeed, Rescher sometimes uses the expression 'truth-criterion' (e.g., 1977, 81), which may suggest that satisfaction of a rule suffices to establish the truth of a proposition. But I do not think that Rescher takes satisfaction of a rule to be criterial for truth in the sense that it either constitutes or demonstrates truth. Rather, satisfaction of a rule provides a warrant or justification for acceptance of the proposition as true (cf. 1977, 79-80).

belief may sometimes lead to success, but it could hardly be supposed to do so on a routine basis:

> Isolated successes can be gratuitous and probatively impotent, but the situation will be otherwise when what is at issue is not isolated actions based on particular beliefs, but a general *policy* of acting, based on a generic and methodologically universalized standard of belief-validation. When one views man as a vulnerable creature in close interaction with a hostile (or at best neutral) environment, it is – to be sure – conceivable that action on a false belief or even set of beliefs might be successful, but it surpasses the bounds of credibility to suppose that this might occur systematically, on a wholesale rather than retail basis. Given a suitable framework of metaphysical assumptions, it is effectively impossible that success should crown the products of *systematically* error-producing cognitive procedures. (Rescher, 1977, 89-90)

Here, in a manner resembling the 'no miracles' version of the success argument for realism, Rescher dismisses the idea of a pragmatically successful but systematically erroneous method as incredible. The crucial factor is the rational implementation of belief in what Rescher describes as a 'highly reactive environment' (1977, 84), 'a duly responsive nature' that is 'complex and volatile' (1977, 91). In such a world, a method of belief-formation that regularly gives rise to successful practical action cannot, in Rescher's words, be 'systematically error-producing'. Quite the contrary, it must surely be 'truth-indicative'.

I shall delve no further into the intricacies of Rescher's methodological pragmatism, though pertinent questions might usefully be raised regarding the line of reasoning that underlies the proposed metaphysical rationale for the truth-indicativeness of method.[15] The purpose of my discussion of Rescher is simply to illustrate how metaphysical considerations may be brought to bear on the problem of method and truth. To further illustrate this, I will now turn to Kornblith's account of the ground of inductive inference.

Kornblith's Natural Ground of Induction

As we saw in 5.3, Hilary Kornblith proposes a naturalistic account of the reliability of induction. The account combines psychologically informed epistemology with a realist metaphysics of natural kinds. Kornblith takes epistemology to be directed to two questions: '(1) What is the world that we may know it?; and (2) What are we that we may know the world?' (1993a, 2). His reply is that mind and world fit

---

15 It is, however, important to note two issues to which Rescher's approach immediately gives rise. The first is the apparent circularity involved in drawing upon substantive principles about the world in arguing that methods of inquiry yield truths about the world. Rescher admits the circularity. Instead of being vicious, however, he seeks to show that the justification of method by practice is cyclical and self-supporting (1977, Ch. 7). The second is the nature of the reasoning from metaphysical principles to the truth-indicativeness of inquiry procedures. In our (2000, 51), Robert Nola and I assimilate the reasoning involved to inference to the best explanation. However, Rescher resists this interpretation (private communication). He argues that it is instead an inference to best systematization. (See Rescher, 2001, Ch. 10 for comparison of inference to the best explanation with inference to the best systematization).

together. On the one hand, properties which occur together in natural kinds make reliable induction possible. On the other hand, our minds are naturally equipped with a conceptual and inferential apparatus tuned to the natural kind structure of the world.

Kornblith takes the success of science to show that natural kinds are the ground of induction (1993a, 41-2). Such success is due to the development of theories about the unobservable structures that underlie the observable properties of things. The classifications devised on the basis of such theories reflect real divisions between natural kinds of things, rather than merely nominal or interest-relative kinds.

> Inductive inferences can only work, short of divine intervention, if there is something in nature binding together the properties which we use to identify kinds. Our inductive inferences in science have worked remarkably well, and, moreover, we have succeeded in identifying the ways in which the observable properties which draw kinds to our attention are bound together in nature. In light of these successes, we can hardly go on to doubt the existence of the very kinds which serve to explain how such successes were even possible. (1993a, 42)

Thus, Kornblith argues that the reliable use of induction in science can only be explained by means of real natural kinds which support induction. It is only if the properties of a member of a kind form a union on the basis of which they must co-occur that induction which projects such properties to unobserved members of the kind could possibly succeed on a reliable basis.

To complete the fit between mind and reality, Kornblith argues that the human mind is disposed to form concepts and draw inferences in ways that reflect real natural kinds. However, I shall not discuss this issue here, since my aim in discussing Kornblith is to draw attention to the role of metaphysics in dealing with the problem of method and truth. Kornblith explains the reliability of induction on the basis of real kinds in nature. It is because members of a natural kind share properties in common with other members of the kind that our inductions about the properties of members of the kind prove to be reliable. Thus, Kornblith employs facts about the nature of reality to explain why induction is reliable. He therefore employs metaphysical considerations to explain why use of a method of inquiry leads to truth.

The Moral of the Metaphysical Story

The approaches of Rescher and Kornblith represent two contrasting approaches to the problem of method and truth. Rescher argues that success in practical application reveals the truth-indicative character of rules of method. Kornblith takes successful use of induction to require the existence of real kinds in nature which make reliable induction possible. Rescher emphasizes the practical implementation of method, while Kornblith draws on empirical research. Rescher's approach forms part of a general theory of the nature and justification of method, whereas Kornblith's account is restricted to the reliable use of induction.

But, despite the contrasts, the approaches of Rescher and Kornblith are united by a deeper commonality. For both approaches exemplify a synthetic solution to the problem of method and truth, which employs metaphysical considerations to establish a connection between method and truth. Both Rescher and Kornblith appeal to the success of science and action in order to argue that our methods provide epistemic warrant with respect to the truth of our beliefs and theories. Both approaches locate the success of method within a broader metaphysical framework which involves assumptions about the nature of the world we inhabit as well as about ourselves as actors and inquirers. Moreover, the metaphysical assumptions employed by both approaches are all broadly consonant with realism.[16]

The latter point deserves emphasis. In their attempt to connect method with truth, both Rescher and Kornblith deploy metaphysical assumptions that are broadly realistic in spirit. Such assumptions cannot therefore be rejected by the realist on metaphysical grounds. The question is whether such metaphysical assumptions should be allowed to play the epistemological role which Rescher and Kornblith ascribe to them. Yet it is entirely unclear how to solve the problem of method and truth in the absence of metaphysical assumptions. I therefore see no alternative but to put the realist's metaphysical assumptions to epistemological use in a manner such as that illustrated by Rescher and Kornblith.

8.5 Conclusion

In this chapter I have sought to raise the problem of method and truth as a challenge to epistemological aspects of Alan Musgrave's scientific realism. The chapter has been largely an exercise in comparative epistemology, which examines alternative solutions to the problem. In line with Musgrave's analysis of the inherent idealism of internal realism, I have argued that the internal realist solution to the problem is not available to the scientific realist. I have also sought to show that Musgrave's own appeal to strictly epistemic principles fails to provide a satisfactory solution to the problem, since such principles do not preclude anti-realist forms of theory-acceptance. By contrast, I have attempted to show that metaphysical considerations are necessary in order to explain why satisfaction of methodological rules warrants acceptance of a theory as true. In this final section, I seek to extract relevant lessons from my analysis with respect to the epistemology of Musgrave's scientific realism.

In the first place, as a realist, Musgrave should have no particular cause to baulk at metaphysical assumptions of the sort described in the previous section. For example, the metaphysical principles introduced by Rescher in relation to human agency, causal interaction and the nonconspiratorial nature of reality, seem entirely acceptable to the realist.[17] The principles are compatible with a realist commitment to

---

16 That the metaphysical considerations to which Rescher and Kornblith appeal are broadly consonant with realism is perhaps most tellingly illustrated by noting that both of their approaches are compatible with a metaphysical realist commitment to an objective, mind-independent reality. Rescher adopts a general principle of uniformity of nature, while Kornblith opts for a somewhat more substantive metaphysics of natural kinds. But both the commitment to the uniformity of nature and to the reality of natural kinds are entirely consonant with a metaphysical realist commitment to mind-independence.

17 As for the natural kinds to which Kornblith appeals in his account of induction, here the realist might have cause to object either to the particular account of natural kinds that

an objective, mind-independent reality. At base, they are no more than an articulation of a commonsense view of ourselves, our surroundings and our relationship to those surroundings. And, in Musgrave's view at least, the scientific realist is not just a realist about science but a realist about common sense as well.

But Musgrave might baulk at appeal to the uniformity of nature. The reason would not be his realism, though, but his anti-inductivism (see Musgrave, 1993, Ch. 9). Here Musgrave's realism must simply be played off against his anti-inductivism. For what is it to be a scientific realist, if it is not to say that there is a real world in which observed phenomena are caused in a lawlike manner by the action of unobservable entities? Of course, we might wrongly identify the causal processes and laws of nature which govern the phenomena. Or the world might be radically transformed overnight. But these are merely sceptical points. The world that we inhabit is a world of objectively existing things, real causal relations and law-governed phenomena. Such a world is characterized by underlying natural uniformities which it is the business of science to discover. A realism that denies this is realism in name only. Indeed, it is realism without the real world.

In section 8.3, I objected to Musgrave's epistemic principle of best explanation that nothing prevents the adoption of an anti-realist analogue of the principle. Yet, as we have seen (8.4), metaphysical resolution of the problem of method and truth proceeds by way of inference to the best explanation of success (or similar form of inference). It might appear inconsistent to object to inference to best explanation in one context while seemingly embracing it in another. My point, however, is not that the realist may do without an explanatory pattern of inference altogether. Given the gap between method and truth, some form of explanatory reasoning must play a role in the epistemology of scientific realism. My point, rather, is that inference to the best explanation as such is not the exclusive domain of the realist. The anti-realist may take it to be justified to accept the best explanation but decline to accept it as true in the realist sense.

However, a realist outcome may be secured once explanatory inference is set within a broadly realist metaphysical framework. In the spirit of the approaches considered in 8.4, I submit that the problem of method and truth is best dealt with as follows. We begin with common sense. The world of common sense is an independently existing reality of causally interacting objects which may or may not be observable by us. We employ a variety of methods to inquire into the ways of this world. Some methods are purely observational, while others are rules of theory appraisal. On the whole, our sense experience provides us with true beliefs about the observable world. In addition, our theoretical reasoning about unobservable states of affairs is frequently rewarded with success at the level of observation and practical action. Given the sort of world we inhabit, the best explanation of the systematically successful implementation of a method of inquiry is that the method provides a reliable means of discovery of truth about the world. Like us, our methods are

Kornblith employs or to the existence of natural kinds, as such. But the idea that there is a real world, in which there are real, non-conventional differences between different sorts of things, is not something to which any realist should seriously wish to raise objections.

fallible. But in a world such as ours the use of such methods could not consistently meet with success, if they were not for the most part a reliable guide to the truth.

In 8.3, I also argued that critical rationalism does not explain why survival of criticism warrants truth as the unique mode of theory acceptance. Yet I do not oppose the method of criticism as such. Indeed, I take the method of criticism to be largely constitutive of the methodology of science. For, as pointed out previously, both falsificationist rules of empirical test and non-falsificationist criteria of theory appraisal may serve as the basis of the critical method in science. The question is simply one of why a theory which survives criticism need be accepted as true.

As with the previous point, this question becomes manageable if the critical method is placed within a broader metaphysical context. If a theory is subjected to a battery of demanding tests, consistently yielding accurate predictions in a range of different circumstances, such performance under test is to be accorded evidential weight with regard to the truth of the theory. It is true, of course, that occasional predictive success may occur as the result of good fortune or accident. But in the sort of world that we inhabit pervasive error is not rewarded by systematic success. A theory which survives a range of rigorous tests may ultimately fail as a result of deeper and more detailed investigation. But in order to sustain systematic success across a great variety of tests, it must either contain a considerable portion of truth or approximate the truth sufficiently closely for it to be empirically indistinguishable from the truth.

It might, finally, be objected that appeal to metaphysical considerations in an epistemological context must proceed in a circle. In order for a claim about reality to justify a method of inquiry there must be reason to accept the claim about reality. But there can be no reason to accept a claim about reality until some method of inquiry is justified.

Such circularity is surely to be avoided. But to insist that epistemology proceed without metaphysics is to fail to appreciate the task with which the realist is confronted. It is not just that the methods of inquiry must be shown to be rationally justified. Since the purpose of inquiry is to discover truth, the methods must be shown to promote the search for truth. But since truth is a matter of how the world is, it must be shown that the methods lead to truth about a mind-independent world. But this requires that something substantive be said about the nature of the world in virtue of which the world is accessible to our methods of inquiry.

The ultimate aim of such an account is a coherent structure in which claims about methods and claims about reality fit together in relations of mutual support. To suppose that such relations of mutual support must result in circular justification is to mistake the nature of epistemology. For human knowledge is a natural phenomenon like any other. To explain how humans know the world requires that we explain how human inquirers may be related to reality in such a way that they may know it. Thus metaphysics and epistemology go hand in hand. For the realist, at least, facts about reality must be brought to bear on facts about inquiry if we are to explain how inquiry yields truth about reality.

Bibliography

Armstrong, David M. (1983), *What is a Law of Nature?*, Cambridge University Press, Cambridge.

Armstrong, David M. (1995), 'What Makes Induction Rational?', in J. Misiek (ed.), *The Problem of Rationality in Science and its Philosophy*, Kluwer Academic Publishers, Dordrecht, pp. 45-54.

Armstrong, D. M. (1999), 'A Naturalist Program: Epistemology and Ontology', *Proceedings and Addresses of the American Philosophical Association* 73, 77-89.

Boyd, Richard (1984), 'The Current Status of Scientific Realism', in Leplin (1984), pp. 41-82.

Boyd, Richard (1991), 'Realism, Anti-Foundationalism and the Enthusiasm for Natural Kinds', *Philosophical Studies* 61, 121-48.

Brown, Harold I. (1988), *Rationality*, Routledge, London.

Campbell, Keith (1988), 'Philosophy and Common Sense', *Philosophy* 63, 161-74.

Carnap, Rudolf (1936), 'Testability and Meaning', in R. Grandy (ed.), *Theories and Observation in Science*, Prentice-Hall, Englewood Cliffs, NJ, 1970, pp. 27-46.

Carnap, Rudolf (1956), 'The Methodological Character of Theoretical Concepts', in H. Feigl and M. Scriven (eds), *Minnesota Studies in Philosophy of Science, Vol. 1*, University of Minnesota Press, Minneapolis, pp. 33-76.

Carrier, M., Roggenhofer, J., Küppers, G. and Blanchard, P. (2004) (eds), *Knowledge and the World: Challenges Beyond the Science Wars*, Springer, Berlin and Heidelberg.

Chakravartty, Anjan (2007), *A Metaphysics for Scientific Realism: Knowing the Unobservable*, Cambridge University Press, Cambridge.

Chalmers, Alan F. (1982), *What Is This Thing Called Science?* (2nd ed.), University of Queensland Press, St. Lucia.

Chalmers, Alan F. (1990), *Science and its Fabrication*, Open University Press, Milton Keynes.

Churchland, Paul (1979), *Scientific Realism and the Plasticity of Mind*, Cambridge University Press, Cambridge.

Coffa, Alberto (1991), *The Semantic Tradition from Kant to Carnap: To the Vienna Station*, Cambridge University Press, Cambridge.

Devitt, Michael (1981), *Designation*, Columbia University Press, New York.

Devitt, Michael (1991), *Realism and Truth* (2nd ed.), Blackwell, Oxford.

Devitt, Michael (2001), 'Incommensurability and the Priority of Metaphysics', in Hoyningen-Huene and Sankey (2001), pp. 143-57.

Devitt, Michael (2002), 'A Naturalistic Defence of Realism', in Marsonet (2002), 12-34.

Devitt, Michael and Kim Sterelny (1987), *Language and Reality*, Blackwell, Oxford.

Dewey, John (1986), *Logic: The Theory of Inquiry*, in *John Dewey: The Later Works,*

Vol. 12, 1928, Southern Illinois University Press, Carbondale.

Ellis, Brian (1985), 'What Science Aims to Do', in P. Churchland and C.A. Hooker (eds), *Images of Science*, University of Chicago Press, Chicago, pp. 48-74.

Ellis, Brian (1990), *Truth and Objectivity*, Blackwell, Oxford.

Ellis, Brian (1999), 'Causal Powers and Laws of Nature', in H. Sankey (ed.), *Causation and Laws of Nature, Australasian Studies in History and Philosophy of Science, Volume 14*, Kluwer Academic Publishers, Dordrecht, pp. 19-34.

Ellis, Brian (2001), *Scientific Essentialism*, Cambridge University Press, Cambridge.

Enç, Berent (1976), 'Reference of Theoretical Terms', *Noûs* 10, 261-82.

English, Jane (1978), 'Partial Interpretation and Meaning Change', *The Journal of Philosophy* LXXV, 57-76.

Feyerabend, Paul (1975), *Against Method*, New Left Books, London.

Feyerabend, Paul (1978), *Science in a Free Society*, New Left Books, London.

Feyerabend, Paul (1981), *Realism, Rationalism and Scientific Method: Philosophical Papers, Vol. 1*, Cambridge University Press, Cambridge.

Field, Hartry (1972), 'Tarski's Theory of Truth', *Journal of Philosophy* 69, 347-75.

Fine, Arthur (1975), 'How to Compare Theories: Reference and Change', *Noûs* 9, 17-32.

Friedman, Michael (1993), 'Remarks on the History of Science and the History of Philosophy', in Horwich (1993), pp. 37-54.

Ghins, Michel (2005), 'Howard Sankey on Scientific Realism and the God's Eye Point of View', *Epistemologia* XXVIII, 139-50.

Giere, Ronald N. (1989), 'Scientific Rationality as Instrumental Rationality', *Studies in History and Philosophy of Science* 20, 377-84.

Goldman, Alvin I. (1986), *Epistemology and Cognition*, Harvard University Press, Cambridge, Mass.

Goodman, Nelson (1955), *Fact, Fiction and Forecast*, Harvard University Press, Cambridge, Mass.

Hacking, Ian (1983), *Representing and Intervening*, Cambridge University Press, Cambridge.

Hanson, N.R. (1958), *Patterns of Discovery*, Cambridge University Press, Cambridge.

Harman, Gilbert (1965), 'Inference to the Best Explanation', *Philosophical Review* 74, 88-95.

Hempel, Carl G. (1965), *Aspects of Scientific Explanation*, The Free Press, New York.

Hooker, C.A. (1987), *A Realistic Theory of Science*, SUNY Press, Albany, NY.

Horwich, Paul (1990), *Truth*, MIT Press, Cambridge, Mass.

Horwich, Paul (ed.) (1993), *World Changes: Thomas Kuhn and the Nature of Science*, MIT Press, Cambridge, Mass.

Horwich, Paul (ed.) (1994), *Theories of Truth*, Dartmouth, Aldershot.

Hoyningen-Huene, Paul (1993), *Reconstructing Scientific Revolutions: Thomas S. Kuhn's Philosophy of Science*, University of Chicago Press, Chicago.

Hoyningen-Huene, Paul, Oberheim, Eric and Anderson, Hanne (1996), 'On

Incommensurability', *Studies in History and Philosophy of Science* 27, 131-41.

Hoyningen-Huene, Paul and Sankey, Howard (eds), (2001), *Incommensurability and Related Matters: Boston Studies in the Philosophy of Science, Vol. 216*, Kluwer Academic Publishers, Dordrecht.

Kitcher, Philip (1978), 'Theories, Theorists and Theoretical Change', *The Philosophical Review* 87, 519-47.

Kitcher, Philip (1993), *The Advancement of Science*, Oxford University Press, Oxford

Kornblith, Hilary (1993a), *Inductive Inference and its Natural Ground*, MIT Press, Cambridge, Mass.

Kornblith, Hilary (1993b), 'Epistemic Normativity', *Synthese* 94, 357-76.

Kornblith, Hilary (2002), *Knowledge and its Place in Nature*, Oxford University Press, Oxford.

Kripke, Saul (1980), *Naming and Necessity*, Blackwell, Oxford.

Kuhn, Thomas S. (1970a), *The Structure of Scientific Revolutions* (2nd ed.), University of Chicago Press, Chicago.

Kuhn, Thomas S. (1970b), 'Reflections on my Critics' in Lakatos and Musgrave (1970), pp. 230-78.

Kuhn, Thomas S. (1977), *The Essential Tension*, University of Chicago Press, Chicago.

Kuhn, Thomas S. (2000), *The Road Since STRUCTURE*, J. Conant and J. Haugeland (eds), University of Chicago Press, Chicago.

Lacey, Hugh (1997), 'The Constitutive Values of Science', *Principia* 1, 3-40.

Lakatos, Imre (1970), 'Falsification and the Methodology of Scientific Research Programmes', in Lakatos and Musgrave (1970), pp. 91-196.

Lakatos, Imre (1974), 'Popper on Demarcation and Induction', in Schilpp (1974), pp. 241-73.

Lakatos, Imre and Musgrave, Alan (eds) (1970), *Criticism and the Growth of Knowledge*, Cambridge University Press, Cambridge.

Lange, Marc (2004), 'Would Direct Realism Resolve the Classical Problem of Induction?', *Noûs* 38, 197-232.

Laudan, Larry (1977), *Progress and its Problems*, University of California Press, Berkeley.

Laudan, Larry (1984), *Science and Values*, University of California Press, Berkeley.

Laudan, Larry (1996), *Beyond Positivism and Relativism*, Westview Press, Boulder, CO.

Lehrer, Keith (1990), *Theory of Knowledge*, Routledge, London.

Leplin, Jarrett (ed.) (1984), *Scientific Realism*, University of California Press, Berkeley.

Lycan, William (1988), *Judgement and Justification*, Cambridge University Press, Cambridge.

Lyons, Timothy D. (2006), 'Scientific Realism and the Strategema de Divide et Impera', *British Journal for the Philosophy of Science* 57, 537-60.

Marsonet, Michele (ed.) (2002), *The Problem of Realism*, Ashgate, Aldershot.

Marsonet, Michele (2002), 'The Limits of Realism', in Marsonet (2002), pp. 190-204.

McMullin, Ernan (1984), 'A Case for Scientific Realism', in Leplin (1984), pp. 8-40.

McMullin, Ernan (1987), 'Explanatory Success and the Truth of Theory', in N. Rescher (ed.), *Scientific Inquiry in Philosophical Perspective*, University Press of America, Lanham, pp. 51-73.

Musgrave, Alan (1988), 'The Ultimate Argument for Scientific Realism', in Nola (1988), pp. 229-52.

Musgrave, Alan (1993), *Common Sense, Science and Scepticism*, Cambridge University Press, Cambridge.

Musgrave, Alan (1996), 'Realism, Truth and Objectivity', in R.S. Cohen, R. Hilpinen and Qiu Renzong (eds), *Realism and Anti-Realism in the Philosophy of Science*, Kluwer Academic Publishers, Dordrecht, pp. 19-44.

Musgrave, Alan (1999), *Essays on Realism and Rationalism*, Editions Rodopi, Amsterdam and Atlanta.

Musgrave, Alan (2001), 'Metaphysical Realism versus Word-Magic', in D. Aleksandrowicz and H. G. Russ (eds), *Realismus Disziplin Interdisziplinaritat*, Editions Rodopi, Amsterdam and Atlanta, pp. 29-54.

Nagel, Ernest (1961), *The Structure of Science*, Routledge and Kegan Paul, London.

Newton-Smith, W.H. (1981), *The Rationality of Science*, Routledge and Kegan Paul, London.

Nola, Robert (1987), 'The Status of Popper's Theory of Scientific Method', *British Journal for the Philosophy of Science* 38, 441-80.

Nola, Robert (ed.) (1988), *Relativism and Realism in Science*, Kluwer Academic Publishers, Dordrecht.

Nola, Robert (1999), 'On the Possibility of a Scientific Theory of Scientific Method', *Science and Education* 8, 1-13.

Nola, Robert and Sankey, Howard (2000), 'A Selective Survey of Theories of Scientific Method', in R. Nola and H. Sankey (eds), *After Popper, Kuhn and Feyerabend: Recent Issues in Theories of Scientific Method, Australasian Studies in History and Philosophy of Science, Volume 15*, Kluwer Academic Publishers, Dordrecht, pp. 1-65.

Nola, Robert and Sankey, Howard (2007), *Theories of Scientific Method: An Introduction*, Acumen Publishing, Chesham.

Papineau, David (1992), 'Reliabilism, Induction and Scepticism', *Philosophical Quarterly* 42, 1-20

Parrini, Paolo (1998), *Knowledge and Reality: An Essay in Positive Philosophy*, Kluwer Academic Publishers, Dordrecht.

Popper, Karl R. (1959), *The Logic of Scientific Discovery*, Unwin Hyman, London.

Popper, Karl (1972), *Objective Knowledge*, Oxford University Press, Oxford.

Popper, Karl (1974a), 'Lakatos on the Equal Status of Newton's and Freud's Theories', in Schilpp (1974), pp. 999-1013.

Popper, Karl (1974b), 'Ayer on Empiricism and Against Verisimilitude', in Schilpp (1974), pp. 1100-1114.

Psillos, Stathis (1997), 'Naturalism Without Truth?', *Studies in History and Philosophy of Science* 28, 699-713.

Psillos, Stathis (1999), *Scientific Realism: How Science Tracks Truth*, Routledge, London.

Putnam, Hilary (1975a), *Mathematics, Matter and Method: Philosophical Papers, Volume I*, Cambridge University Press, Cambridge.

Putnam, Hilary (1975b), *Mind, Language and Reality: Philosophical Papers, Volume II*, Cambridge University Press, Cambridge.

Putnam, Hilary (1978), *Meaning and the Moral Sciences*, Routledge and Kegan Paul, London.

Putnam, Hilary (1981), *Reason, Truth and History*, Cambridge University Press, Cambridge.

Quine, W.V.O. (1969), *Ontological Relativity and Other Essays*, Columbia University Press, New York.

Quine, W.V.O. and Ullian, Joseph (1970), *The Web of Belief*, Random House, New York.

Reisch, George (1991), 'Did Kuhn Kill Logical Empiricism?', *Philosophy of Science* 58, 264-77.

Rescher, Nicholas (1977), *Methodological Pragmatism*, Basil Blackwell, Oxford.

Rescher, Nicholas (1982), *Empirical Inquiry*, Rowman and Littlefield, Totowa, NJ.

Rescher, Nicholas (2001), *Philosophical Reasoning: A Study in the Methodology of Philosophizing*, Blackwell Publishers, Oxford.

Resnik, David (1994), 'Hacking's Experimental Realism', *Canadian Journal of Philosophy* 24, 395-412.

Ristau, Carolyn (1991), 'Aspects of the Cognitive Ethology of an Injury-Feigning Bird, the Piping Plover', in Ristau (ed.), *Cognitive Ethology: The Minds of Other Animals*, Erlbaum, Hillsdale, New Jersey, pp. 91-126.

Salmon, Wesley (1984), *Scientific Explanation and the Causal Structure of the World*, Princeton University Press, Princeton.

Sankey, Howard (1991), 'Translation Failure Between Theories', *Studies in History and Philosophy of Science* 22, 223-36

Sankey, Howard (1994), *The Incommensurability Thesis*, Avebury, Aldershot.

Sankey, Howard (1997), *Rationality, Relativism and Incommensurability*, Ashgate, Aldershot.

Sankey, Howard (2004), Comment on Holm Tetens, 'Scientific Objectivity with a Human Face', in Carrier, Roggenhofer, Küppers and Blanchard (2004), pp. 95-8.

Sankey, Howard and Hoyningen-Huene, Paul (1996), 'Interview: Paul Hoyningen-Huene', *Metascience* 10, 59-70

Sankey, Howard and Hoyningen-Huene, Paul (2001), 'Introduction' in Hoyningen-Huene and Sankey (2001), pp. vii-xxxiv.

Scheffler, Israel (1967), *Science and Subjectivity*, Bobbs-Merrill, Indianapolis.

Schilpp, Paul A. (ed.) (1974), *The Philosophy of Karl Popper: Library of Living Philosophers, Volume 14*, Open Court, La Salle, Illinois.

Schlick, Moritz (1938), 'Meaning and Verification', in O. Hanfling (ed.), *Essential Readings in Logical Positivism*, Blackwell, Oxford, 1981, pp. 32-43.

Sellars, Wilfrid (1963), *Science, Perception and Reality*, Routledge and Kegan Paul, London.

Smart, J.J.C. (1963), *Philosophy and Scientific Realism*, Routledge and Kegan Paul, London.

Stanford, Kyle (2003), 'No Refuge for Realism: Selective Confirmation and the

History of Science', *Philosophy of Science* 70, 913-25.

Sterelny, Kim (1983), 'Natural Kind Terms', *Pacific Philosophical Quarterly* 64, 110-125

Stich, Stephen, (1990), *The Fragmentation of Reason*, MIT Press, Cambridge, Mass.

Tarski, Alfred (1943), 'The Semantic Conception of Truth', in Horwich (1994), pp. 107-41.

Tetens, Holm (2004), 'Scientific Objectivity with a Human Face: Four Reflections from a Pragmatist Point of View', in Carrier, Roggenhofer, Küppers and Blanchard (2004), pp. 81-93.

Thagard, Paul (1978), 'The Best Explanation: Criteria for Theory Choice', *The Journal of Philosophy*, 75, 76-92.

Van Fraassen, Bas (1980), *The Scientific Image*, Oxford University Press, Oxford.

Wilkerson, T.E. (1995), *Natural Kinds*, Avebury, Aldershot.

Worrall, John (1988), 'The Value of a Fixed Methodology', *British Journal for the Philosophy of Science* 39, 263-75.

Worrall, John (1989), 'Fix It and Be Damned: A Reply to Laudan', *British Journal for the Philosophy of Science* 40, 376-88.

Index